EXPERIMENTING WITH HUMANS

AND ANIMALS

JOHNS HOPKINS

INTRODUCTORY STUDIES

IN THE HISTORY

OF SCIENCE

Mott T. Greene
and Sharon Kingsland,
Series Editors

Experimenting with Humans and Animals

From Galen to Animal Rights

Anita Guerrini

THE JOHNS HOPKINS UNIVERSITY PRESS

BALTIMORE

The Johns Hopkins University Press
2715 North Charles Street
Baltimore, Maryland 21218-4363
www.press.jhu.edu

Library of Congress Cataloging-in-Publication Data

Guerrini, Anita, 1953–
Experimenting with humans and animals : from Galen to animal rights / Anita Guerrini.
    p.   cm. — (Johns Hopkins introductory studies in the history of science)
    Includes bibliographical references and index.
    ISBN 0-8018-7196-4 (hc) — ISBN 0-8018-7197-2 (pbk)
    1. Vivisection—History. 2. Animal experimentation—History. 3. Human experimentation in medicine—History. I. Title. II. Series.
QP45 .G84 2003
619′.09—dc21   2002011064

A catalog record for this book is available from the British Library.

In memory of
Richard S. Westfall
1924–1996

Contents

Preface

When I first began teaching a course on the history of animal experimentation in 1992, I found that no general work discussed human and animal experimentation from its beginnings in antiquity to the present. The leading histories of physiological research, Thomas Hall's *History of General Physiology* and Elizabeth Gasking's *Rise of Experimental Biology,* were nearly thirty years old. It seemed time to reexamine some of the issues they raised and to explore others not yet examined. I decided, therefore, to write a history of animal and human experimentation. After all, the scientific and ethical ideas of René Descartes and of the physiologists of the nineteenth century did not spring from their heads like Athena from the head of Zeus. Rather, the ideas and attitudes that have led in the West to a biology and medicine reliant on experiments on humans and animals date from the very beginnings of Western thought. Today, the utility and ethics of animal and human experimentation are passionately debated. I am convinced of the value of history as a means to understanding differing points of view, and I hope that this work is a small contribution toward a dialogue among the opposing parties.

The history of science is blessed with a multiplicity of methods, approaches, and viewpoints. Ethicists have other avenues of approach, as do sociologists and anthropologists. I have used aspects of many of these approaches, but I have not attempted to introduce, or even acknowledge, all of them in this book; such a cacophony of voices would make the book unreadable. I list some of the more noteworthy texts in Suggested Further Reading.

The late French philosopher Michel Foucault greatly influenced historians with his emphasis on the relationship between power and knowledge. Certainly the relationship between humans and the rest of nature, particularly animals, is based on power. A large number of recent works explore this idea in terms of race, class, or gender, but fewer have looked at animal-human relationships through a Foucauldian lens. The human body has been much studied of late as a site of power relationships, focusing mainly on sex or gender. A few recent studies of early modern anatomy have gone beyond sex to look at the body as a subject and object of culture. In addition, some feminist theorists have appraised animal experimentation. Perhaps the most innovative work has come from sociologists and those historians who use sociological methods. The concept of "boundary objects," first described by Susan Leigh Star and James Griesemer, can be fruitfully applied to both human and animal experimental subjects. Boundary objects overlap intersecting social worlds and realms of knowledge. Similarly, human and animal subjects held different meanings for different audiences; candidates for smallpox inoculation in the eighteenth century, for example, were medical patients, experimental subjects, statistical units, and members of distinct social groups all at the same time. Exciting work is now being done on modern laboratory life and the role of animals and animal models, and I acknowledge some of this work in the conclusion.

I am a historian, not an ethicist, and my focus in this book is on practices as much as on ideas. This book makes no claim of providing a complete or conventional history of animal and human experimentation, which would occupy a volume many times its size. It is, rather, a subjective look at what I consider to be significant and telling events in that history in the Western world. It is episodic rather than forming a continuous narrative. I am very conscious of what I have omitted, and I regret particularly that I had neither expertise nor space to discuss non-Western ideas and practices. I have tried to present a balanced account, but my own biases are probably evident. Let me state at the outset that I believe animal and human experimentation is a necessary part of modern science and medicine, and that it will not disappear soon. But I do not think experimentation on living subjects should be taken lightly or used indiscriminately. It should be regulated, alternatives should be thoroughly investigated, and living subjects should receive the best possible treatment and care.

Of course abuses have occurred, and still do. But my purpose in writing this history is not simply to reveal abuses or to revel in the triumphs of science. I wish the story were that straightforward, because it would have been much easier to write. The story instead is one of trial and error, prejudice and leaps of faith, clashing egos and budget battles. It is too easy to take an Olympian perspective and designate certain historical players as right and others as wrong, certain ideas as winners and others as losers. As a historian of science, I have spent much of my career examining "losing" ideas, not only for their intrinsic interest as intellectual problems, but for what they reveal about human personalities and events. In addition, a close examination of almost any historical idea or problem reveals unforeseen complications and contexts. Few things are simply right or wrong, either ethically or scientifically. More often they are a muddle of mixed motives and half-clear ideas.

The overriding theme of this book is not the conflict between science and ethics, as might be expected, but the interaction between science and society. Someone with a different philosophical or disciplinary approach would, no doubt, look at this theme differently. My argument is that the values of science are the values of the society it inhabits. Science is a privileged enterprise only insofar as society allows it to be.

In writing this book, I have relied upon the work and knowledge of many other people. A number of these people are listed in the notes and the recommended readings, but I would like to acknowledge a few others. Lawrence Badash first suggested I write a book on this subject. Paul Farber and members of the History of Science colloquium at Oregon State University shredded an early version of my chapter outline and forced me to think about what I really wanted to say. Paul has been a sounding board throughout the writing of this book. The series editors, Mott T. Greene and especially Sharon Kingsland, also gave much good advice. Brian Martin shared with me quantities of material on the AIDS-polio connection. A chance remark by Ann Carmichael on the number of monkeys used in polio research led me to focus a chapter on polio.

I have imposed on many friends and colleagues to read portions of this book, including Teresa Burgess, John Gluck, Peter McDermott, Marcia Meldrum, Karen Rader, Lindsey Reed, Jo-Ann Shelton, and Peter Sobol, and I am grateful for their time and effort. They are of course in no way responsible for bizarre opinions or errors of fact. Marcia Meldrum and Karen Rader shared unpublished material with me, and Andrea Rusnock sent material on smallpox. A number of people helped me find illustrations. Participants on the H-Sci-Med-Tech list, the C18-L list, and the H-Albion list answered all kinds of questions, and librarians at the University of California, Santa Barbara; the University of California, Los Angeles; and the University of Southern California helped find books, articles, and illustrations.

The penultimate draft of this book was written during the course of a fellowship at the Centre Alexandre Koyré in Paris. I am grateful to Dominique Pestre and Nadine Dardenne at the Centre Koyré, and to John Krige at the Centre de recherche en histoire des sciences et des techniques at La Villette for facilitating my work. Robert J. Brugger at the Johns Hopkins University Press has been far more patient with me than I deserved, and his assistant, Melody Herr, has been a delight to work with.

Students at the University of California, Santa Barbara, between 1992 and 1998 who took my class on the history of animal experimentation deserve special recognition for their willingness to ask hard questions and not always accept my answers. Peter Neushul, then a graduate student, was a fantastic research assistant for the first year of that class and brought to my attention many materials I would never otherwise have found. Our ongoing conversations on this topic continue to enlighten me. I am grateful to the University of California, Santa Barbara, for an Instructional Improvement Mini-Grant in 1992 that helped pay for his assistance.

Members of the research focus group in Human-Animal Relationships at the University of California, Santa Barbara, have probably heard more about this book than they ever wanted to hear. They have been an unending source of inspiration and support. I would also like to thank my fellow members of the Institutional Animal Care and Use Committee at the University of California, Santa Barbara, especially Diane McClure, who have taught me a lot about how regulation and oversight really work. Most of all, I want to thank my family. Michael Osborne, my husband, colleague, and best critic, as usual read many drafts and gave trenchant advice. He and our sons, Paul and Henry Osborne, also gave me their patience and love.

EXPERIMENTING WITH HUMANS

AND ANIMALS

Introduction

What responsibility do we humans bear for things we do to other people and other creatures in the name of scientific advancement? Few questions in contemporary life arouse deeper feelings. Animal rights activists, believing deeply in the justice of their cause, break into laboratories, destroy research, and threaten scientists. The sense of public outrage escalates when reports surface, as they have done in recent years, that the American government once conducted experiments—exposure to atomic radiation, infection with disease, subjection to LSD—on unsuspecting citizens. The proper place of people and animals in scientific practice raises important and instructive issues concerning the nature of scientific investigation, the ethical treatment of laboratory subjects, and the relationship between science and society.

This book seeks to introduce such issues by means of historical episodes, each of which rewards close inspection. Taken together, they demonstrate the development of a central idea in Western thought: challenging nature in order to reveal the truth. These stories thus also sketch the broad history of experimental biology, physiology, and medicine.

Although confidence in the experiment dates from Western antiquity, the meaning of *experiment* depends on time and place. We might more accurately describe the work of the ancients as *demonstrations* rather than experiments; rather than aiming to discover new facts about nature, they were more often intended to illustrate received beliefs. Today we say that an experiment tests a hypothesis, a question about nature, under controlled conditions. A researcher or investigator begins with the hypothesis and designs the experiment so that it answers the question truthfully, that is, according to how nature really acts and not necessarily as the observer would want it to act. The investigator controls the setting of the experiment in order to account strictly for all the factors at play and to be reasonably certain of what cause in the experiment led to what effect.

Experimenters therefore monitor physical conditions such as temperature and air quality, use precise instruments, and carefully select subjects. In many biological experiments and clinical trials, researchers deliberately do not administer the experimental procedure or substance to a "control group" of subjects, to be monitored for purposes of comparison. Instead, administering placebos—pills or treatments that do not contain the experimental element—performs the same function. In clinical trials, the experiment is controlled for any possible psychological bias through the use of "double-blind" methods in which neither the persons administering substances nor the persons receiving them are aware of which are the experimental substances. Note that obviously failed experiments, in which nature did not act as the hypothesis predicted, can be as valuable as successful ones.

This approach to the experiment owes much to practices, findings, and ideas of scientific professionalism that date from the late nineteenth and twentieth centuries. Although as early as the eighteenth century statistics helped to make the case for smallpox inoculation, statistical techniques became increasingly important in the twentieth century as researchers sought to minimize the influence of their preferences and also take into account the detailed characteristics of the subjects they studied. A contemporary experiment may employ subjects chosen completely at random or test the same remedy on several groups of subjects in a sequence of "block" patterns. Experiment yields information, or data, which in recent experience usually appears in the form of quantitative statements of number, proportion, or statistical significance, and that information is then further manipulated with the use of statistical techniques to make explicit the experimental aims.

Although observers often refer to experimentation on living beings (especially on animals) as "vivisection," that term, strictly defined, means the cutting open of the animal body. Of course, many experiments do not involve vivisection. In the pages that follow, the distinction between vivisection and other forms of experimentation will be made as clearly as possible, but historically, commentators did not always employ precise definitions.

Attitudes toward animals and toward experimentation began to develop in antiquity with Aristotle, the Alexandrians, and the Roman physician Galen, who established dissection (the opening up of a dead body) and vivisection as methods to learn about the inner workings of the human and animal body. Dissection could show essential structures, but only vivisection, argued the Alexandrians and Galen, could reveal the function of these structures. Early Christian theologians, following Aristotle, believed that while humans existed for their own sakes (possessed of reason and immortal souls), animals (lacking reason) existed to serve human needs. These ideas of the relative place of humans and animals in nature—and therefore about experimentation—prevailed in the West for centuries.

In the seventeenth century a dedicated English researcher and physician named William Harvey conducted a series of experiments over a period of twenty-five years searching for answers about the motion of the blood in animals and humans. Harvey relied heavily on animal dissection and vivisection in reaching his conclusion that the blood circulated throughout the body. His experimental methods, which included the use of inductive reasoning (where many specific examples lead to a statement of general principles), became the basis of all further experimental work on animal and human function. In sharp contrast to Harvey's use of inductive reasoning, the contemporary French philosopher René Descartes used deductive reasoning (going from the fundamental or general principle [that he was a thinking being] to the specific) to declare that body and soul were separate entities and that living bodies acted like machines. Descartes argued that because animals lacked reason, they could not feel pain in the same way humans did. Most experimenters did not believe this, and philosophical debate over animal experimentation continued in the eighteenth century. In 1789, the English jurist and philosopher Jeremy Bentham redefined the animal-human relationship with his much-quoted statement, "the question is not, Can they *reason*? nor, Can they *talk*? but, Can they *suffer*?"

Bentham placed animals potentially within the realm of human ethics, but he hardly resolved the debate. Mid-nineteenth-century developments in physiology produced the first organized movement in opposition to experimenting on animals. The growth of experimental science occurred mainly in continental Europe, especially France and Germany, while the animal protection movement was for many years centered in Britain. The introduction of anesthesia in the 1840s complicated experimental issues. Should experimenters relieve pain in subjects? If so, under what circumstances?

The early-eighteenth-century controversy over smallpox inoculation raised other issues, mostly concerning human subjects, that remain with us today. Some physicians at the time claimed that inoculation—giving a mild version of the disease to a potential victim—conferred immunity to smallpox. But no one could explain how or why. Inoculation therefore remained experimental, and risky. Many people, greatly fearing smallpox, accepted the risks of being inoculated. Comparing survival rates between those persons who were inoculated and those who were not argued on behalf of the procedure. Yet the basic question of how much risk was acceptable remained unanswered.

By the end of the nineteenth century, with scientists and antivivisectionists increasingly polarized, new knowledge in the growing field of microbiology, including the germ theory of disease, seemed to support the cause of science. The work of the French biologist Louis Pasteur on rabies and the development of arsphenamine, a yellow crystal powder that fought syphilis and trench mouth, won adherents to "scientific progress." Even so, some lay observers continued to doubt the value of vaccines, the wisdom of uncontrolled scientific experiments, and the continued use of animals in research.

Unquestionably the ugliest examples of systematic and predatory use of both animals and human subjects in physiological research took place in Nazi Germany before and during World War II, when captive men and women underwent torture at the hands of people wearing white lab coats. Because Nazi experimentation on humans in particular so grossly violated any reasonable standard of morality, the episode of Nazi "science" more properly belongs in studies of atrocities than in this one, and yet it helps to make several points. First, while Nazi experimentation on humans represented an extreme that we hope humans will never revisit, experimentation without consent went on elsewhere at the time and has happened since. Second, the Nuremberg Code, which has formed the basis for all discussion and regulation of human experimentation since World War II, developed as a direct result of the trials of Nazi war criminals at Nuremberg and as a response to the extremes of Nazi research. We must remember that efforts since then to regulate and provide oversight of scientific research on human subjects originated with the Nazi horrors. Finally, Nazi research raises a less obvious but necessary ethical question: Should researchers employ data obtained from inhumane experiments?

Taking the large view, medical and scientific research has made the lives of many if not all people far better than their lives otherwise would have been. We are the heirs, and in many ways the beneficiaries, of the legacy of Western science. Nonetheless, medical, scientific, and technological breakthroughs bring with them ever more difficult moral questions. In the twentieth century, disease research used far more animals than physiological research, and yet the treatment these animals received tended to win little notice because researchers generally used less-valued animals, mainly mice and rats. Post–World War II polio research offered a case study both in the methods of modern biomedical research, including the first example of a large-scale controlled human clinical trial, and in the laboratory use of animals much more like humans—primates. Indeed, the use of monkeys and apes in research has become increasingly controversial, partly because of the possibility of monkey diseases transferring to humans and partly because studies of primate intelligence and behavior have revealed close similarities to humans. Primates may make ideal research subjects, but are they enough like humans to make drastic intervention into their lives (depriving young monkeys of their mothers, for example) or to make subjecting them to vivisection seem unnecessarily cruel and inhumane? Some activists have argued that apes should be granted the same moral status as humans, excluding them from experimental use. Should we draft one set of moral guidelines for primates and another for rodents?

These and other ethical questions about animal and human use in research remain with us, unanswered and vexing. Debate has gone on with little knowledge of the historical background of animal and human experimentation. Historians believe, and this book may demonstrate, that understanding the roots of serious problems aids in our ability to deal with them. To understand where we are going we need to understand from where we have come.

1 Bodies of Evidence

Experimentation and Philosophical Debate in Premodern Europe

Alexandria in the year 280 B.C.E. was a young city, vibrant and restless. Its polyglot inhabitants had migrated there from every corner of Alexander the Great's now-shattered empire. Alexander had founded the city in 332 B.C.E. as a model of Greek culture, as a civilized, cosmopolitan society that valued science and the arts. It was that, and more. Its academy, library, and medical school became dominant forces in Western culture for a thousand years. In Alexandria, Greek culture thinly overlay a multitude of languages, religions, and ideas. For a time, people were willing to put aside old taboos and try out new ideas and practices.

Nonetheless, murmurs must have arisen in the marketplace at the news that the Greek physician Herophilus (ca. 330–ca. 260 B.C.E.) and his younger colleague Erasistratus (304–245 B.C.E.) had been given permission by the king to cut into a living man, not to cure him but merely to look inside. The man was a condemned criminal, scheduled for execution. Arguments for and against were heated. No contemporary accounts survive, but we hear faint echoes of these arguments in *De medicina,* the account written three centuries later by the Roman historian Celsus.

Many medical sects flourished in the wide-open atmosphere of Alexandria, but the opinions held by the followers of two of these, the empirics and the dogmatists, are especially relevant. The empirical medics who set up practice in stalls in the marketplace were strongly opposed to any kind of dissection. Unlike the "philosophical" physicians who declared that knowledge of the natural world and how it worked was essential to medicine, the empirics, as they were known, opposed any kind of theory. They relied on observation of the patient, claiming that philosophy and experimentation were irrelevant to medical practice. The physician could gain knowledge of disease only through this repeated observation. The great physician Hippocrates, in the casebooks known as the *Epidemics,* provided ample support to this view, with page after page of symptoms and signs and very little theorizing. Dissection and vivisection, said the empirics, gave knowledge of the dead, not of the living. In vivisection, they argued, the subject died anyway in the course of the operation, and the act of vivisection itself caused pathological changes that brought the validity of the observations into question. Observation of wounds during the course of treating them could give much the same information without deliberately injuring a fellow human being. The empirics objected to vivisection on intellectual grounds and also on moral ones, because the physician, as healer, should not cause suffering and death.

According to the historian Celsus, Herophilus and Erasistratus were members of another sect known as the dogmatists. Dogmatists believed that a knowledge of anatomy was critical to medical practice. They dissected to learn more about the body's internal workings, because mere observation of its exterior, even if supplemented by glimpses of wounds, was inadequate. In this they followed the model of the Greek philosopher and naturalist Aristotle (384–322 B.C.E.), who had been Alexander's tutor. Although Aristotle's goal was to learn more about how the human body works, he had only limited access to the human body itself. Strong taboos against the mutilation of the human body existed in many ancient cultures, including ancient Greece. Surgery in the modern, invasive sense was unknown in antiquity—both because of these taboos and because of the inherent risks of infection and shock—and was confined to wound treatment and bone setting. Elaborate embalming and mummification rituals in ancient Egypt involved opening the dead body and removing its organs, but the Egyptian physician was not allowed to cut open his patients. In Greece, mutilation of the dead body was regarded with horror, reserved for one's direst enemies in the course of war. In the ancient tragedy *Antigone,* the title character goes to great lengths to ensure the proper burial of her brother's body. For Aristotle and the dogmatists then, dissection and experimentation on animals was a necessary alternative to research on humans.

Herophilus and Erasistratus practiced animal dissection and vivisection, but they believed that, though useful, it was insufficient. Dissection of human cadavers was also necessary. In Alexandria, Egyptian embalming practices may have helped sanction this activity, but not the next step. The Alexandrian physicians agreed with Aristotle that dissection of the dead, while it tells us much about the structure of the body, tells us nothing of living function. There is a fundamental distinction, said Aristotle, between life and death. The logical conclusion of this reasoning would lead to dissection of live humans. Celsus pointed out that deliberate vivisection gave more valuable information than chance observation of wounds, which would be in a pathological rather than a normal state. He offered the argument that the sacrifice of a few for the many was justified. Celsus lived and wrote in ancient Rome, where a criminal could "pay" for his crime by making his body useful to the community.

For empirics and dogmatists, moral arguments based on the undoubted cruelty of vivisection—either of animals or of humans—were much less important than intellectual arguments about the best method of obtaining knowledge about the body. In a later period, early Christian commentators would cite human vivisection in Alexandria as an example of the depravity of pagans. In 280 B.C.E., human lives—particularly the lives of criminals, slaves, or enemies in war—were not highly valued. Slaves were regularly tortured in courts of law in order to extract evidence, and the mutilation of slaves while they were alive indicated their lower status. But the ancient taboos against the mutilation of the body did not take into consideration the notion of pain or cruelty. As we shall see, changing definitions of pain and its significance play a major role in changing concepts of animal and human experimentation.

The Alexandrians derived much of the philosophical basis for their work from the writings and example of Aristotle. Therefore let us leave Alexandria for a time and travel to Athens, half a century earlier. In 347 B.C.E., the philosopher Plato died in Athens at the age of eighty. It was widely expected that Aristotle, his star pupil, who had studied with the master for two decades, would succeed him as the head of his school, the Academy. But Plato did not choose Aristotle to succeed him. Aristotle was not an Athenian, and it was soon evident that he disagreed with Plato on several fundamental issues. Aristotle left Athens and embarked on a career that would change the course of Western thought.

Plato believed that the external world was but an imperfect imitation of the ideal "forms," or ideas, which could only be perceived intellectually. Aristotle, on the contrary, believed that the form of an object could not be separated from its manifestation in nature, thereby establishing the study of the external world as an important and worthwhile enterprise. Plato's philosophy centered on humans, but Aristotle looked at nature as a whole and viewed the human as another animal, although certainly the most highly developed of all. "In all natural things," wrote Aristotle, "there is something wonderful." Even ugly and insignificant animals displayed the creative power of nature and were worthy of study. Most of all, nature revealed an underlying purpose, what Aristotle called the "for something's sake,"* which led to knowledge of the beautiful, the ultimate goal of Plato's thought. This notion of purpose in nature, known

---

*Aristotle, *De partibus animalium*, trans. J. Balme (Oxford: Clarendon Press, 1972), 645a 15–20; 20–25.

Human Vivisection

The ancient Roman historian Celsus described the arguments for and against human vivisection. The medieval *wound-man's* injuries and weapons were intended to aid the memory of the surgeon, and demonstrate the Empirici's argument that much could be learned of internal organs in the course of treating injuries.

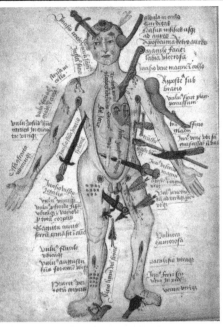

[The dogmatists] hold that Herophilus and Erasistratus did this in the best way by far, when they laid open men whilst alive—criminals received out of prison from the king—and whilst these were still breathing, observed parts which beforehand nature had concealed. . . . For when pain occurs internally, neither is it possible for one to learn what hurts the patient, unless he has acquainted himself with the position of each organ or intestine . . . nor is it, as most people say, cruel that in the execution of criminals, and but a few of them, we should seek remedies for innocent people of all future ages.

On the other hand, those who are called "Empirici," because they have experience . . . contend that inquiry about obscure causes and natural actions is superfluous, because nature is not to be comprehended. . . . But what remains, [it is] cruel as well, to cut into the belly and chest of men whilst still alive, and to impose upon the Art which presides over human safety someone's death, and that too in the most atrocious way. Especially this is true when, of things which are sought for with so much violence, some can be learnt not at all, others can be learnt even without a crime.

■ Quote from Cornelius Celsus, *De medicina*, trans. W. G. Spencer, Loeb Classical Library (Cambridge: Harvard University Press, 1960), lines 23–27, 40–43. Illustration: Wound-man, fifteenth century, Wellcome MS 290, f. 53 verso. Courtesy of Wellcome Trust Medical Photographic Library, London.

as *teleology* (from *telos,* the Greek word for purpose or meaning in the sense of fulfillment or completion), has proven to be a powerful and enduring idea in Western thought.

To Aristotle, the study of animal structure and function was an important means of learning about nature. To this end, he dissected many dead animals, at times killing an animal specifically for that purpose. In addition, he experimented on (but did not cut open) live animals, arguing that living function could be explored only in the living. Few others had investigated nature in this way. Animals had been part of religious rituals for millennia, and earlier Greek writers had observed animals and had written about them. But Aristotle's interest in their internal structure and function was new.

Aristotle admitted that dissection could be unpleasant. The act of cutting, and the look, smell, and feel of the internal parts of the body, all were distasteful, and few people were willing to attempt this practice. Dissection, then and now, is also a difficult skill to master, requiring patience, a strong stomach, and manual dexterity. The vascular system, for example, was difficult to discover in dead animals because the vessels collapsed after death. In living animals, the veins were usually not visible to an external observer. The best solution, according to Aristotle, was to starve an animal and then strangle it. Looking at old men had taught him that blood vessels stand out more clearly in an emaciated body, and strangling prevented blood loss.

Following Hippocrates, Aristotle assumed an analogy existed between human and animal, especially mammalian, structures. Humans were another animal. But Aristotle never considered dissecting a human, and it would not have been allowed in Greece in his time. Nonetheless, knowledge of the human body was his ultimate goal. To him, the overriding purpose of nature was the fulfillment of the chain of being, which defined a hierarchical arrangement of nature with humans on top and animals below, descending down a ladder or chain to plants, all the way down to rocks. To Aristotle, each species achieved its proper form and realized its proper potential, but on a scale. Everything had a place. The hierarchy of nature mirrored the hierarchy of society, in which men were superior to women, and some humans naturally ruled while others were natural slaves. Rationality was the key to both hierarchies, for the mind always dominated the body.

But if humans were at the top of the scale because of their superior minds, how could they also be analogous to lower animals? The analogy existed entirely in the body. The nature of the scale gave a clue, for the spaces between species were not large and abrupt, but small and incremental. Many aspects of animal anatomy were quite similar to that of humans, as was obvious even from external observation, and animals could legitimately act as stand-ins to examine the inner parts of humans.

A fundamental difference nonetheless separated humans from animals. Aristotle believed that only humans had intelligence and therefore rational souls; animal souls possessed emotion but not reason. Humans and animals therefore did not occupy the same moral plane. Because animals were not rational and were incapable of deliberate choice, he concluded that there is no such thing as justice or injustice toward animals. Aristotle never spoke about the morality of using animal bodies, either alive or dead, and the use of animals for sacrifice, food, and labor similarly were not moral issues for him. Such notions would have seemed absurd, as indeed did the notion that all humans, whether male or female, free or slave, citizen or noncitizen, had certain rights.

Theophrastus (ca. 372–287 B.C.E.), Aristotle's successor as head of the Lyceum, his school in Athens, took his master's arguments about analogy a step further, arguing that the physical analogy between humans and animals had deeper implications. In contrast to Aristotle, Theophrastus believed animal sacrifice displeased the gods. He argued that vivisection and meat eating were inhumane: if we feel pain, so do animals. Humans and animals share a kinship based on physical and mental similarities; therefore, to deprive an animal of life was an unjust act. Modern ethicists still debate whether loss of life, as well as suffering, is a harm and therefore morally unjust. But Theophrastus tempered his position in practice by the equally compelling notion of necessity. Sometimes, he said in his treatise "On Piety," it is necessary to kill animals to eat, and this is permissible. Even if one deplores the spilling of blood, it is sometimes necessary for a society to execute criminals. Only unnecessary cruelty is to be avoided. By this argument of necessity, vivisection could fall under the category of necessary cruelty and even human vivisection might also be permissible.

At about the time that Theophrastus wrote, Herophilus and Erasistratus in Alexandria required no philosophical justification for vivisecting humans or animals. Although many later authors condemned them, some, like the Roman physician Galen (ca. 130–210 C.E.) envied them. In fact, much of what we know of the methods and discoveries of the Alexandrians (whose own works did not survive) comes from Galen's writings.

Galen quoted extensively from Herophilus's treatise *On Anatomy.* Herophilus was the first to distinguish nerves from other tissues, and he went on to establish that the brain rather than the heart (as Aristotle had argued) was the center of the nervous system, and that the perception of pain depended on both brain and nerves. His descriptions of the brain—incorporating work on human and ox brains—far surpassed those of Aristotle in accuracy and detail. He also distinguished between sensory and motor nerves. His work on the eye revealed the existence of the optic nerve as well as the retina, which he named. Many of the names Herophilus gave to anatomical features have survived, such as the "styloid process" of the skull, so called for its similarity to a long, slender writing implement known as a stylus. He also offered the first detailed description of the human liver and compared it to the animal liver. The liver was of great importance in ancient Greek medical theory as the supposed source of blood, and in religion, divination from the inspection of the liver of a sacrificial animal was a common practice. Moreover, Herophilus clearly described a human liver, not an animal liver as Aristotle had, and his description must have been a result of numerous dissections of humans. Herophilus also offered a detailed account of the vascular system, distinguishing between arteries and veins, which Aristotle had not clearly demarcated.

Herophilus's younger contemporary Erasistratus also worked on the vascular system. His interest was in function as much as in form, and he performed many experiments on live animals, especially pigs, oxen, and goats. He explained the role of valves in the heart, and compared the action of the heart to a pump or bellows. These observations were independently rediscovered in the seventeenth century C.E. by William Harvey. Erasistratus distinguished three kinds of vessels in the body: veins, arteries, and nerves (which he believed to be hollow). Each of these vessels contained a different fluid or gas: veins contained blood, nerves contained a fluid that conveyed impulses to the brain, and arteries contained air.

How could Erasistratus argue that arteries contained air? Weren't the facts obvious? In dead bodies, he saw empty arteries, because in death the blood moves to the veins as it exhausts its supply of oxygen. But in live bodies, blood comes from an artery when it is cut. Erasistratus explained that blood rushes into a cut artery from the veins because the air in the artery rushes out, creating a vacuum. He reconciled two apparently contradictory observations with a theory that explained both. Later researchers explained this same experimental result differently.

Herophilus and Erasistratus represent the high point of Alexandrian medical research. Human vivisection and dissection in antiquity began and ended with them, and after them animal experimentation also fell into disuse for a time. The empirics won the battle of the medical sects, and both Greek and Egyptian taboos against mutilation of the body reasserted themselves. Later kings in Alexandria were less receptive to Greek scientific ideas. In premodern societies, the cultural sanction for scientific activity was often fragile and fleeting. Galen alone, some four hundred years after the Alexandrian physicians, again reached their level of expertise, and many of their findings were not confirmed until the seventeenth century, when conditions for research were comparable.

Galen, Gladiators, and Apes

Galen, physician to Emperor Marcus Aurelius, had declared in his tract "that blood is contained in the arteries" (*An in arteriis natura sanguis contineatur*), not air. A follower of the ideas of Erasistratus issued a public challenge and attempted several public demonstrations to prove his point, and Galen in turn accepted the challenge, experimenting on a number of live animals. In the most telling experiment, Galen cut open an animal to expose a length of artery, tied off the artery in two places, and cut the artery between the ligatures. Blood, not air, poured from the cut. The audience, no doubt, applauded.

There was indeed an audience, for neither Galen nor his challenger were about to resolve a public challenge in private, and anatomy was one of many forms of entertainment on the crowded streets of ancient Rome. Galen had already participated in the public dissection of an elephant, and bested his rivals by demonstrating the existence of a bone in its heart. In the highly competitive arena of Roman medicine, reputations could be made or broken with the stroke of a knife. (Galen had left Rome under a cloud in 166 C.E., only to return triumphantly with the emperor three years later.) The rivalry of public dissection was the medical equivalent of gladiatorial combat, a practice with which Galen was quite familiar, since he had spent the early years of his medical career as a physician and surgeon at the school for gladiators in his native town of Pergamum in Asia Minor.

Galen was, with Hippocrates, the most important physician in antiquity, and his ideas dominated Western thought until the Renaissance. Along with the Alexandrians, he was one of the greatest anatomists and experimenters in antiquity. The rediscovery in the sixteenth century of his manual for dissection, *On Anatomical Procedures,* introduced his methods of dissection and animal experimentation to a new generation of researchers and established these methods as the only legitimate means of obtaining knowledge of anatomy and physiology. Without Galen, Harvey could not have made his discoveries.

Galen spent much of his early life as a peripatetic student of medicine, visiting many sites in the Empire, including Alexandria, where he saw a complete human skeleton for the first and only time in his life. Human dissection had long ceased to be performed there, but Alexandria's medical school remained the most famous in the Western world. Along the way, Galen mastered Greek philosophy and attained impressive skills in anatomy. He perfected the latter as the gladiators' surgeon, learning anatomy, as the empirics advised, by observing wounds.

But in his many books Galen severely criticized the empirics, aligning himself instead with the dogmatic school. The physician, he said in his *On Anatomical Procedures,* was not a mere craftsman but a man of learning. Knowledge of inner structures and functions was essential to successful medical practice, and direct experience was the best teacher. Yet Galen did not disdain book learning, and he was as much a synthesizer as an original researcher; often the only surviving evidence of the writings of earlier practitioners is to be found in Galen's works.

His works number twenty hefty modern volumes, and that is not his entire production. The works which most concern the topic of animal and human experimentation are *On Anatomical Procedures,* which is the text of lectures on anatomy he delivered in the year 177, and *On the Usefulness of the Parts of the Body,* a general description of the human body written about 175. Both of these works were originally written in Greek.

Galen believed that all living beings have *physis,* or life. In addition, animals possess what he called *psyche,* or consciousness. Humans possess both these qualities, but they also have the capacity to reason. Like Aristotle, Galen sought evidence of purpose in nature. He argued that each part of the body has a purpose, and it was the job of the anatomist to determine that purpose. The body as a whole fit into the overall plan of the great chain of being. Galen aimed, in *On the Usefulness of the Parts of the Body,* to provide a complete description of the human body, and the work is a remarkable performance. But he never dissected a human or even witnessed such an event. While he made use of the work of his predecessors, especially the Alexandrians, he based many of his conclusions on animal dissection and vivisection.

In Galen's Rome, human dissection was simply not allowed. The naturalist Pliny the Elder (23–79 C.E.) wrote disapprovingly of those who employed human body parts for folk cures, such as scraping sore gums with the tooth of a man who died violently. In his writings, Galen wistfully mentioned those long-ago days in Alexandria when human bodies could be opened to the physician's gaze. In his day, the best he could do was to dissect a Barbary ape, perhaps with a naked slave nearby to serve as a comparison. The analogy between humans and animals, based on anatomical similarities, was of course a central tenet of the great chain of being, and provided another guarantee of the regularity and planning evident in the universe. Galen's extensive use of animals reinforced and strengthened the analogy between animals and humans, making a virtue of necessity and firmly establishing animal experimentation as the standard method for learning about human anatomy and physiology.

In *On the Usefulness of the Parts,* Galen offered a complete system of anatomy and physiology, organized around the central philosophical principle of teleology: the "usefulness" of the parts was not merely their function, but their purpose, what Aristotle called the "for something's sake." Both this principle and his reliance on animal anatomy led Galen to some erroneous conclusions about human anatomy, which critics in the Renaissance later pointed out.

Medical theory, at this point in history, in its symmetry and correspondences, demonstrated the meaningfulness of nature as a whole. It was based on Aristotle's concept of four elements (fire, air, water, and earth) and four qualities (hot, wet, cold, and dry). In Galenic medical theory, four humors—phlegm, blood, black bile, and yellow bile or choler—governed health by their surfeit or lack. Each humor derived from a particular organ—the brain, the heart, the spleen, and the gall bladder, respectively—and governed a particular temperament. "Phlegmatic," "sanguine," "melancholy," and "choleric" still describe certain mental states, even though we no longer believe in the theory of the humors. When a person was in perfect health, the humors existed in equitable quantities and in harmony with one another, but an imbalance would cause illness. Because each person's natural balance and temperament differed, so each experience of illness was individual. Generalizations about disease, which the empirics hoped to gain, were therefore of limited value. Repeated encounters with patients, such as those described in the *Epidemics* of Hippocrates, could help the physician to identify salient signs, but treatment of each case had to be carefully tailored to the individual. In addition, many other factors, including weather, geography, and astronomical phenomena, might contribute to the illness.

Therapies, therefore, concentrated on restoring balance, often by administering medicines or treatments opposing the effect of the excessive humor. So colds, caused obviously by excessive phlegm (a cold, moist humor), responded to hot, dry medicines. One sign of fever was a flushed face, caused by an excess of hot, wet blood. A common therapy was to let blood out of the patient, restoring the proper balance, and the patient did become cooler and pale as a result. Diet and bathing (hot or cold) were also important therapies.

The physiology that Galen detailed in *On the Usefulness of the Parts* supported this theory of illness and therapy. He viewed digestion as a critical function that helped to establish the balance of the humors; too much of the wrong kind of food could cause all manner of ill health. His description of the process by which digested food became blood emphasized the separation from the blood of the two forms of bile and their proper placement in the spleen and gall bladder. He also spent much effort on describing the process by which wastes were eliminated from the body, another important aspect of therapy. In addition, the detailed anatomical descriptions of limbs, and of bone, skin, and tendon had direct application to the treatment of wounds, and Galen spoke harshly to those who would ignore dissection and needlessly injure their patients out of ignorance.

But much of Galen's work in anatomy had little direct application to medical theory or practice, and when he listed the uses of anatomy, medical practice came last. He constantly searched for purpose in nature. He dissected many different kinds of animals, he said, to show that a single mind had conceived them. Equally important was what Galen called "knowledge for its own sake."* Although he declared near the beginning of On Anatomical Procedures that the most important function of anatomy was to enable the physician to treat wounds, this theme quickly receded before the thrill of discovery.

As Aristotle had discovered, dissection was not easy, and in the meticulous descriptions in On Anatomical Procedures we see a master at work. No detail was too small or unimportant for Galen; he advised the diligent dissector to remove the skin of the animal himself, for careless assistants could damage important structures. This attentiveness in dissection trained the student for the more difficult task of vivisection. Galen's accounts of vivisection do not make for easy reading. His matter-of-factness, his brash delight in besting his intellectual enemies, and the clinical detail of his accounts still disturb readers two millennia later. But in Galen's work, we can see the beginnings of the Western tradition of biological research, and his own utter absorption in his task draws the reader along, however unwillingly.

Although therapeutic benefit received an occasional mention—and the research described in On Anatomical Procedures formed the basis for his therapeutics in On the Usefulness of the Parts—there is no mistaking Galen's passion for research as an end in itself. When he described opening the chest of a living animal to view its beating heart, the hemorrhage from cut arteries was simply an annoyance; when the thorax was cut in a certain way, he noted, the animal ceased to breathe or cry out, but if the anatomist covered the cuts with a hand, breathing and crying out resumed. If certain nerves were cut or damaged, the animal would lose its voice entirely, or if merely half-cut, the animal would lose only some of its voice. His interest was in the effects, not in the animal, which was just a tool, a means to an end.

The nervous system, and the effects of cutting certain nerves, fascinated Galen. He cut the spinal cord in pigs, goats, and apes, noting the degree of paralysis produced. He also noted which functions were controlled by which nerves. He then went on to study brain function: which part of the brain con-

* Galen, On Anatomical Procedures, trans. Charles Singer (London: Oxford University Press, 1956), 33–34.

Dissecting a Pig

Galen dissects a pig in this illustration from a sixteenth-century edition of Galen's works. He described his method of experimentation in this passage from *On the Natural Faculties:*

The fact is that those who are enslaved to their sects are not merely devoid of all sound knowledge, but they will not even stop to learn! Instead of listening, as they ought, to the reason why liquid can enter the bladder through the ureters, but is unable to go back again the same way—instead of admiring Nature's artistic skill—they refuse to learn; they even go so far as to scoff, and maintain that the kidneys, as well as many other things, have been made by Nature *for no purpose!* . . . We were, therefore, further compelled to show them in a still living animal the urine plainly running out through the ureters into the bladder; even thus we hardly hoped to check their nonsensical talk.

Now the method of demonstration is as follows. [Galen then described his procedure on a living animal, tying off the ureters to demonstrate that they pass urine from the kidneys to the bladder. The ureters were then untied and the bladder was allowed to fill, and the penis was tied off to prevent urination. The urine, however, remained in the bladder, even when it was squeezed, and did not return to the kidneys.] Now, if anyone will test this for himself on an animal, I think he will strongly condemn Asclepiades [Galen's rival], and if he also learns the reason why nothing regurgitates from the bladder into the ureters, I think he will be persuaded by this also of the forethought and art shown by Nature in relation to animals.

■ Quote from Galen, *On the Natural Faculties,* I.13, trans. A. J. Brock, in *Source Book of Greek Science,* ed. M. R. Cohen and I. E. Drabkin (Cambridge: Harvard University Press, 1948), 481–82. Illustration: Galen, *Opera omnia* (Rome: Giunti, 1596–97). Courtesy of Biomedical Library, University of California, Los Angeles.

trolled which part of the body? Could he convince those who doubted that the brain and not the heart controlled the nervous system? Dissecting ox brains given to him by the butcher was practice for the main event: the vivisection of the brain of an ape, in which he stimulated various parts of the living brain and noted the effects produced on the body.

Galen advised his students to cut "without pity or compassion" into a living animal. He noted that the cries of an animal in pain were part of the procedure, and the "unpleasing expression of the ape when it is being vivisected" was unavoidable.* Galen had little concern for the animals, and by our standards, he was very cruel. But he lived in a world of cruelty and violence in which gladiatorial combat was as common as football, animals were tormented and killed in arenas for amusement, and one form of public execution was to throw condemned humans together with wild animals into the amphitheater for an afternoon's entertainment.

Because animals ranked far below humans in the scale of being, Galen followed Aristotle in granting to animals only limited consciousness, which implied considerably less consciousness of pain. An animal whose thorax had been opened, said Galen, was perfectly unimpaired in its functions after the operation was ended. He compared its state to the case of a slave who had lost part of his sternum yet survived: "it is surely more likely," wrote Galen, "that a non-rational brute, being less sensitive than a human being, will suffer nothing from such a wound."†

Galen discovered more about the animal body, and by analogy, the human body, than had ever been known before. He was the last great original thinker in the Greek tradition, and animal experimentation in his manner was not performed in the West for a thousand years after his death. By chance and circumstance, his works survived when the works of others did not, and he became a major authority in the Christian and the Islamic worlds when his works were translated from their original Greek into Latin and Arabic. He established standards, techniques, and an attitude toward animal experimentation that in many ways still inform the science of today. Galen emerges from his works as a recognizable, if not necessarily an attractive, personality. He forces us to confront the consequences, both good and bad, of animal research.

The Christian Body: Angel or Brute?

Galen seems an unlikely candidate for Christian acceptance. He was not a Christian, and his boastfulness and cruelty have little connection with the biblical "gentle Jesus, meek and mild." Yet Galen's work was accepted by medieval and Renaissance Christians for several reasons: He was deeply religious and celebrated in his work the creativity and miraculous organization of nature, which he attributed to the guidance of an active intelligence, the platonic

---

*Galen, *On Anatomical Procedures, the Later Books,* trans. W. L. H. Duckworth (Cambridge: Cambridge University Press, 1962), 15.
† Galen, *On Anatomical Procedures* (1956), 192.

creator-god. The last book of *On the Usefulness of the Parts* is a paean to the skill of this intelligence, to whom he referred as "Creator" or "Nature" but who could very easily be viewed as God the Creator, of Judeo-Christian scripture. In addition, *On Anatomical Procedures,* the work most revealing of Galen's personality and most explicit regarding his treatment of animals, was not translated from Greek into Latin until the 1530s and was therefore inaccessible to most Western readers after the fall of Rome.

Were Galen's views on animals incompatible with Christian doctrine? What did Christian doctrine have to say about animals? The Christian's relationship to the natural world has been debated since the foundation of Christianity, with implications for modern environmentalism as well as animal welfare. The Bible is by no means straightforward on this issue. In the book of Genesis, God commanded the first humans to "be fruitful and multiply, and fill the earth and subdue it; and have dominion over the fish of the sea and over the birds of the air and over every living thing that moves upon the earth." Later in Genesis, Adam named the animals, conferring upon them identity in terms of his dominance.* But that dominance was not initially one of plunder or indiscriminate consumption. Before the fall, Adam and Eve were vegetarians, and after, God valued the farmer Abel more than the herdsman Cain.

The Hebrew Old Testament includes several passages that contradict the edict to dominate nature. For example, passages in Psalms express an appreciation of, rather than a dominance over, nature. The writer of the book of Ecclesiastes declared, "For the fate of humans and animals is the same; as one dies, so dies the other. They all have the same breath, and humans have no advantage over the animals."† Some early Jewish scholars interpreted such passages to mean that humans were stewards of the earth, entrusted with its care and preservation by God, who alone possessed and ruled it. Because God created humans to complete his creation, each generation should leave the earth more beautiful and fruitful than before. This is an agrarian ideal of a domesticated nature, encompassing Abel the farmer and Cain the herdsman. The ideal nature is tame, not wild.

A more prominent view, held by many early church fathers, argued that because God made humans in his image, unique in their possession of an immortal soul, they were naturally above the rest of nature. This view was compatible with ancient ideas such as the great chain of being, which also placed man at the top of creation. The Greek philosophical school known as the Stoa similarly argued that rationality distinguished humans from the rest of nature

---

*Bible, New Revised Standard Version, Genesis 1:28.
† Ibid., Ecclesiastes 3:19.

and made them closer to the gods. Like the gods, humans therefore enjoyed a natural hegemony over nature, which was made for their use.

Christianity drew from both the Hebrew tradition and from other ancient religious and philosophical traditions. Central to Christian doctrine was its recognition of the value of the individual, whose salvation was the result of his or her behavior in this life. But there was little consensus on what constituted Christian behavior toward animals and the natural world. The New Testament, like the Old, gave no clear guidelines. Jesus employed animals as symbols of faith—the hen gathering her chicks under her wing, the shepherd who rejoices in finding the stray sheep—but he assured his followers that their lives were worth more than those of the beasts.* He transferred demons from humans to pigs and allowed these Gadarene swine to run to their deaths. The evangelist Paul also stated in his first letter to the Corinthians that God's main concern was with humans, not animals. The ultimate concern of Christians was to achieve life after death, and this life was only accessible to humans, who possessed immortal souls.

Many early Christian commentaries emphasized God as creator, and discerned the nature and purpose of God insofar as they detected purpose and meaning in nature, in a manner similar to Galen's in *On the Usefulness of the Parts*. The North African bishop St. Augustine (354–430), the most influential of the early church fathers, developed this theme. In his *City of God*, he declared that God "did not wish the rational being, made in his own image, to have dominion over any but irrational creatures, not man over man, but man over the beasts."† Augustine's views on nature are part of his larger picture of fallen humanity, for nature reflected the consequences of Adam's fall, which changed a perfect landscape into one linked inextricably with death and suffering. Although nature continued to demonstrate God's design and purposes, this design had been corrupted by human sin. The pervasiveness of sin gives a sinister cast to Augustine's introduction of the concept of the "two books," in which the "book of nature" complements the book of God or the Bible. This concept was later used in Christian Europe to defend the practice of science.

Augustine acknowledged the chain of being in his definition of human dominion. Humans, he said, fall between animals and God on the scale, but reason makes them closer to God. While animals are superior to plants, humans are superior to all. Each being on the scale has value as a creation of God, but each is not of equal value. To think that each was of equal value was "the height

---

* Ibid., Matthew 23:37; Luke 15:3–7.

† Augustine, *The City of God*, trans. Henry Bettenson (London: Penguin, 1984), bk. XIX, chap. 15.

of superstition," according to Augustine. He wrote, "there are no common rights between us and the beasts and trees." He added, "We can perceive by their cries that animals die in pain, although we make little of this since the beast, lacking a rational soul, is not related to us by a common nature."* This remained the Christian doctrine for the treatment of animals until modern times.

Two Christians of the Middle Ages, St. Francis of Assisi and St. Thomas Aquinas, revisited Augustine's doctrines in ways that would be important for the experimenters of the sixteenth and seventeenth centuries, who revived the techniques of Galen and the Alexandrians. St. Francis of Assisi (1182–1226) has been praised as the first environmentalist. But a closer look at his work reveals that he viewed animals as part of a human-centered symbolic system. The *Fioretti,* a popular book of stories about Francis, included many instances of his kindness to animals. He preached to the birds, he rescued some doves from a hunter and built them nests, and he persuaded the fierce wolf of Gubbio to stop attacking the townspeople. In "The Canticle of Brother Sun," he praised "Brother Sun" and "Sister Moon." In all of these stories, the larger context is clear. The birds to whom Francis preached demonstrated the creativity of God, and symbolized the Franciscan friars who owned nothing and traveled throughout the world. The wolf of Gubbio served as a reminder to the sinful townspeople of the perils of hell. The doves represented pure souls whom Francis rescued from sin. Brother Sun is a symbol of God himself. An example of the everyday treatment of animals is the story of Brother Juniper, who cut off the foot of a living pig to feed a sick friar. Francis scolded Juniper, not for his cruelty to the pig, but because the pig belonged to someone else.

Francis lived during the High Middle Ages, when Europe was gaining peace and prosperity after the centuries of turmoil that followed the fall of Rome. He could wander around begging without worrying about bandits; he could walk in the woods and preach to the birds without fear. The wolf of Gubbio was a lingering vestige of the old fear of the wilderness, the wild animal leaping into your living room. But as it turned out, he was not very formidable, and like the rest of nature, Christianity and civilization could tame him.

During the lifetime of Francis, an immense project of translation and transmission was under way that reintroduced the works of Aristotle, and some works of Galen, to Western Europe. The impact of Aristotle's philosophy of nature on medieval intellectuals was immediate and overwhelming, and scholars attempted to reconcile Aristotle's ideas about nature with Christian theology. The most important of these scholars was the theologian St. Thomas Aqui-

---

*Augustine, *The Catholic and Manichaean Ways of Life,* trans. D. A. and L. J. Gallagher (Washington, D.C.: Catholic University, 1966), 102, 105.

nas (1225–1274), who delineated the realms of faith and reason in his *Summa theologica*. Because reason was God given, it could lead the faithful to truths compatible with God's truth of revelation. He reiterated Augustine's "two books" notion, agreeing that the book of nature complemented the book of God. Aristotle's teleology supported Thomas's theology: nature was created for a purpose, and that purpose was good. Therefore, nature was good, and its study beneficial. In this period, anatomical studies began to resume in some Italian schools of medicine, and dead animals were the anatomical subjects.

Thomas, like Augustine, also accepted Aristotle's hierarchical view of nature as a chain of being. In the *Summa theologica,* Thomas explained that human superiority was based on the possession of reason, which implied the existence of an immortal soul. Since animals lacked reason, they lacked immortal souls and could not participate in the afterlife. Because they were irrational, they also lacked the ability to choose. Animals appeared to make choices because of the structure of their parts, not because they exercised reason and free will. Thomas compared animals to the new mechanical clocks of the time. If humans can craft such devices, how much more sophisticated are the machines of God? The machine analogy would prove to be long-lived.

Thomas believed that animals were created for the sake of man. Reason tells us, he said, that there are no restrictions on the human use of creatures. On earth, all things exist for human use, to help them to become closer to God. But cruelty can have no role in the godly character. Therefore, although there was no theological stricture against it, good Christians, who were compassionate, would treat animals well. This action did not in itself make humans good, because animals lacked moral status. Goodness was defined solely in terms of behavior toward other humans. But, argued Thomas, cruelty to animals could lead to cruelty toward humans. However, the definition of what constituted cruel behavior varied widely. Thomas's argument became a standard reference for the treatment of animals, and established a theological basis for experimentation on them, although this was not practiced in his time. When experimentation was revived during the Renaissance, justification was readily at hand.

In the ancient and medieval periods in Western culture, Greek concepts about the body and nature joined together with Christian notions about the place of humans in nature and their duties toward the natural world. Together, this powerful combination of ideas formed the framework of the Western point of view toward animals, humans, and experimentation. Although particular ideas were later disputed, the general framework, which declared the superior status of humans among living beings, remained, and influences our ideas even today.

2 Animals, Machines, and Morals

It was a chilly winter day early in 1621. The doctor strode into the room and flung back the shutters, letting in the thin sunlight. The room was cold but he did not notice as he moved rapidly about, setting out tools and instruments. He walked over to a stack of cages and opened the topmost one, gently pulling out a large rabbit. Its nose twitched as the doctor carefully tied it to a board with holes drilled in it through which he passed the thin cord that bound the animal's limbs. The rabbit lay on its back, blinking and quivering, its limbs splayed, its chest rising and falling quickly. The doctor took a sharp, thin-bladed knife and with practiced skill laid open the rabbit's chest. The animal struggled and panted, but the bonds held fast. The doctor sliced through the breastbone and spread open the ribcage with his strong fingers, exposing the rapidly beating heart. He cut a strong thread of silk and tied it around the rabbit's aorta, watching with satisfaction as the animal's heart grew engorged with blood while the vessel beyond the ligature became white. He delicately sliced the aorta and saw the blood spurt out in regular pulses. As the rabbit slowly expired, the doctor seized a notebook and began to write, looking up to observe the rabbit's heart as it slowed.

The doctor, William Harvey (1578–1657), repeated this and other experiments hundreds of times, on dozens of different animals, to prove his theory that the blood circulated through the body. Harvey was the first since Galen to initiate a research program based on experimentation on live animals. His discovery of the circulation of the blood is the most important event in the history of medicine, and marks the beginning of medicine as a science. His research methods, which combined animal dissection and vivisection with experiments and quantitative arguments, formed a model for future research. During the seventeenth century, Harvey's methods were appropriated by natural philosophers who viewed animal and human bodies as machines that could be analyzed in mechanical terms (natural philosophy is the prescientific study of nature, which encompasses both matter and spirit). Although Harvey would not have agreed with this point of view, it resulted in a significant increase in the understanding of the body and its functions. But parallel to this mechanization of the body, and in part in reaction to it, a new sensitivity toward and awareness of animals also began to evolve in this period.

An experiment is consciously devised as a way to try to solve a particular problem. Highly structured, it involves foresight, planning, and the prediction of an outcome (the hypothesis); it shows causes at work in nature, and it entails active intervention in natural processes, not merely passive observation. When Harvey tied a ligature around the rabbit's aorta, he controlled the conditions of the experiment by actively intervening in the normal processes of nature. This intervention was guided by the hypothesis that the blood flowed in one direction and that by blocking its flow certain things would happen. An experiment should be repeatable, with the same results each time, before it can be accepted as a demonstration of how nature actually works. Harvey repeated his animal experiments many times over a period of several years before he was satisfied that what he found was indeed a fact about nature, an accurate observation. He therefore assumed that nature followed certain laws, and that a consistent effect implied a consistent cause. Nature was not arbitrary in its actions. As a Christian, Harvey probably believed that miracles could occur, but they were outside the ordinary course of nature, which in its lawfulness displayed the intelligence of its creator.

By proposing that the blood circulates throughout the body, Harvey advanced a theory that could completely restructure his field of inquiry. He presented a new model for science that entirely changed the rules by which research was to be conducted. Galen believed that the blood did not circulate but moved back and forth in the blood vessels, and that new blood was constantly being made in the liver to replace blood that was used up by the body's organs. This concept of the vascular system dictated the questions other researchers could ask about the blood. According to this explanation, venous blood was new blood as it came from the liver, dark in color and full of nutrients. The lungs served only to cool the warm heart. The Galenic model accommodated new evidence discovered in the sixteenth century by anatomists in Padua, where Harvey studied. In 1543, the young Fleming Andreas Vesalius (1514–1564) had demonstrated that the septum, the muscle dividing the left and right sides of the heart, was solid, and not perforated as Galen had claimed. But he did not go on to question how the blood then got from one side of the heart to the other. A decade later, Realdo Colombo (ca. 1515–1559) concluded from dissection and vivisection that the blood moves from the right side of the heart to the left by passing through the lungs, what we now call the pulmonary circulation. But this was not, to him, sufficient evidence to abandon the Galenic theory, and he continued to believe that the lungs functioned to cool the blood.

To Harvey, the pulmonary circulation was evidence that the Galenic explanation was inadequate, an anomaly that did not fit into the old model. Rather than trying to explain the evidence to fit the old view of the world, he decided that the view of the world was incorrect and that a new theory was necessary to explain the observed phenomena. His experiments helped him to devise his new theory—that the blood circulated and that it was constantly reused rather than constantly newly made—and to test this theory.

Until Harvey challenged the prevailing model, these tests, such as the ligature experiment described at the beginning of the chapter, were not performed because they did not answer questions that were significant to the Galenic explanation. Harvey used other phenomena that had not previously been viewed as anomalies as further proof of his new theory. One phenomenon was the existence of valves in the veins. Harvey's teacher at Padua, Fabricius, thought that valves simply slowed down the flow of blood to extract the maximum nutritive value, following Galen's theory. Harvey, however, believed that the valves kept the blood flowing in one direction, as his theory of circulation required. His experiments followed a modern pattern: one experiment led to another, in a chain or tree. In addition, Harvey developed multiple tests to demonstrate a single experimental fact. He demonstrated the purpose of the valves in the veins not only by inserting a probe into a vein but also by observing the flow of blood in the arm of a human subject with especially prominent veins.

The University of Padua in northern Italy, where Harvey learned anatomy and surgical techniques at the end of the sixteenth century, was the best place in Europe for this kind of study. During the fourteenth and fifteenth centuries, anatomical demonstration for instructional purposes had become commonplace in universities across Europe as old restrictions on the use of dead bodies broke down. These formal demonstrations followed guidelines set down in the early fourteenth century, and included the use of animals, most likely dead animals. Medieval natural philosophy emphasized the authority of the text as the source of knowledge, and anatomical demonstration served to illustrate the text rather than to yield new information. The rediscovery of additional ancient texts in the fifteenth and sixteenth centuries, however, eventually led to major changes in the concept of the body and how we gain knowledge about it.

While medieval scholars discovered classical texts through a complicated process of retrieval and translation, Renaissance scholars sought out original Greek texts and in so doing found many previously undiscovered ones. For example, the medical history of the Roman Celsus was rediscovered in 1478, revealing the many conflicting schools of medical thought in ancient Rome. The rediscovery of many of Galen's works, particularly his *On Anatomical Procedures* in 1531, prompted a revival of animal experimentation. In addition, the passion for realism of Renaissance artists led many of them to attend anatomical demonstrations and even to dissect animals and humans themselves. In his notebooks, Leonardo da Vinci (1452–1519) described dissections and vivisections that he witnessed and performed.

In 1543, Vesalius published his response to Galen, *On the Fabric of the Human Body in Seven Books* (often known from its Latin title as *De fabrica*). Vesalius had been a child prodigy, carving up neighborhood cats and other animals, and he was named professor of anatomy at Padua at the age of twenty-three. He dissected and vivisected for research as well as for classroom demonstration. Vesalius, like other university anatomy lecturers, was granted the bodies of a certain number of executed criminals to use in his annual public lectures, but his research program required many more bodies. At this time, the practice of human dissection was restricted by physical, political, and social circumstances. The difficulty of preserving bodies dictated that dissection take place in winter. Local governments granted the use of executed criminals for public dissection, but were not obligated to provide additional bodies for research purposes. Despite the lack of legal or church restriction on the act of anatomy, ordinary people remained reluctant to volunteer their own bodies or those of their loved ones to be mutilated after death. The professor of anatomy was often called upon to perform autopsies, but this did not allow for extensive research. Anatomists nonetheless tried to persuade families to allow autopsies on their relatives.

Vesalius pursued his passion for research in highly suspect ways. He frequented executions in the hope of obtaining a body fresh from the gibbet. His students held "resurrection parties" to rob new graves. But the supply of bodies, particularly of female bodies (since most executed criminals were males), continued to be inadequate for his purposes. For example, in his research on the female reproductive system between 1537 and 1543, Vesalius had access to only six female bodies. Three of these were for demonstrations, and therefore limited in their usefulness for research; one was the body of a child stolen from its grave, which was too decomposed for much research; one was a murdered woman, on whom Vesalius performed the postmortem. Therefore only one body was available solely for research. Although Vesalius was highly

Showman before a Raucous Crowd

The title page of Vesalius's landmark work *On the Fabric of the Human Body in Seven Books* (1543) depicts Vesalius at center stage, his hands poised in mid-dissection. Like Galen, Vesalius is a showman before a raucous crowd. No professor hovers above the audience reading a text, no assistant does the manual work of cutting. Vesalius was both lecturer and dissector and the equal of the ancients, represented by classically garbed observers who look on respectfully. A dog and a monkey await dissection, or vivisection, in their turn; in a 1540 demonstration, Vesalius used three human bodies, six dogs, and other animals as well.

■ Andreas Vesalius, *De humani corporis fabrica libri septem* (Basel: Johann Oporinus, 1543). Courtesy of Biomedical Library, University of California, Los Angeles.

critical of Galen's reliance on animals to explain human anatomy, it is not surprising that like Galen, Vesalius relied heavily on animals as stand-ins for humans. Vesalius refuted many of Galen's claims, which had been derived from animal anatomy, but the lack of correspondence between animal and human anatomy on some points did not entirely discredit those claims. Vesalius devoted the last chapter of *De fabrica* to the dissection of living animals, agreeing with Galen that function, as opposed to form, could not be fully understood without recourse to living animals.

Vesalius's boastful accounts in *De fabrica* of his methods of obtaining bodies led to the enactment of laws against grave robbing and to fears among many Italians of disrespectful mutilation after death and even accidental (or not so accidental) vivisection. Vesalius omitted some of the more offensive passages from later editions of *De fabrica,* but the damage had been done. Thereafter the anatomist became ever more closely allied with the executioner in the public imagination. In addition, the naturalistically posed, anatomized humans in Vesalius's illustrations presented the human body, much like that of any other animal, as an object for the gaze and knife of the anatomist, blurring the boundary between the animal and the human body. The ancient respect for the human body had weakened considerably.

When Harvey traveled to Padua in 1599, the anatomy chair once occupied by Vesalius was held by Girolamo Fabrizzi, known by his Latin name of Hieronymus Fabricius ab Aquapendente (1537–1619). Fabricius was a devoted Aristotelian, and like Aristotle, he viewed the animal body as a universal concept, built from a number of observations of individual animals, including humans. Aristotle's search for cause and purpose motivated both Fabricius and his pupil Harvey.

Harvey pursued his anatomical researches privately after he returned to London in 1602. His use of animals was dictated by circumstances, since it continued to be difficult to procure human cadavers for private research. But it was also philosophically justified, because like Fabricius, Harvey's goal was to learn about animals in general, and not only about humans. He set out to learn about the motion of the heart, and not initially to prove the idea of circulation, which only gradually occurred to him. Harvey presented some of his conclusions in public anatomy lectures delivered to the London College of Physicians (the physicians' guild, not an educational institution) between 1616 and 1618. As in most anatomy lectures, these included both the dissection of a human cadaver and the dissection and vivisection of several animals. His lectures were very popular, and a critic complained that Harvey did not restrict his audience to the learned: "If only, Harvey, you would not hold an anatomy in front of jacks-in-office, petty lordlings, money-lenders, barbers and such like ignorant rabble, who, standing around open-mouthed, blab that they have seen miracles."* No hint of the circulation emerges in these lectures. Harvey continued to experiment, to make and discard hypotheses. After more than twenty years of work, his observations began to make sense to him in the context of a theory of circulation.

Harvey revealed his theory of circulation in his book *On the Motion of the Heart and Blood in Animals* (known by its Latin title of *De motu cordis*), published in 1628. In his book, he proved three "suppositions," or hypotheses. The first is a quantitative argument: the amount of blood that passes through the arteries is a much greater quantity than the body could possibly produce from a normal amount of nourishment. Harvey computed the amount of blood driven through the heart, using anatomical evidence for the capacity of the heart and estimates of the quantity. Half an ounce of blood, he concluded, was driven out of the human heart at each beat; at two thousand heartbeats an hour, this comes to more than sixty pounds an hour, or nearly fifteen hundred pounds a day. In animals this discrepancy was even more striking, for Harvey calculated that three and one-half pounds of blood pass through the heart of a sheep in half an hour; and as he found by bleeding a sheep, its entire quantity of blood was only four pounds. The body could not use up and constantly make this much blood, so it must be reused.

Harvey's second "supposition" was that the blood is driven to all parts of the body by the beat of the heart and is carried by the arteries. He defined systole or contraction as the active phase of the heartbeat, in contrast to most ancient writers who believed the expansion or diastole of the heart was active. At diastole, Harvey argued, the heart was at rest. The pulse, on the other hand, was caused by the expansion of the arteries as the contracting heart filled them with blood. Harvey's third "supposition," that the blood is returned to the heart by the veins in equal amount as is pumped out to the arteries, logically followed, although he could not observe the connections between the veins and arteries.

Since Harvey sought general principles applicable to all animals, he dissected all sorts of hearts, with two, three, and four chambers, belonging to cold- and warm-blooded animals. The hearts shared certain characteristics: for example, all possessed valves that acted as one-way doors to prevent the back-

*Caspar Hofmann to William Harvey, May 1636, translated in Gweneth Whitteridge, *William Harvey and the Circulation of the Blood* (London: Macdonald, 1971), 241.

ward flow of blood. The valves in the veins operated in the same way. Harvey concluded in a very Aristotelian way that structures such as the valves must have some active physiological purpose.

As we have seen, Harvey tied off veins and arteries to observe the surrounding changes. The left side of the heart became distended when the aorta, the outgoing artery, was tied off. Tying off the vena cava, which sent venous blood into the right side of the heart, caused the vein to become distended and the heart to turn whitish from lack of blood. Harvey also noted that a human arm became cold and blue when the arteries were tied off with a tight tourniquet. Over and over again, Harvey cut an artery near the heart of a living animal and saw the blood forcibly pushed out at each heartbeat. The heart, he announced, acted like a pump.

Harvey employed both ancient and modern methods. He performed experiments to discover new facts about nature and to demonstrate facts that he believed he had already proven, or were generally known. Like Aristotle, Harvey also employed many analogies, reasoning that events or structures similar in some ways would be similar in others. To Aristotle and Harvey, a fundamental analogy existed between human and animal structures; experimenting on animals could yield information about the human body.

Harvey drew another analogy between the macrocosm (big world), or the structure of the heavens, and the microcosm (small world), or the structure of the earth, and between both and the smaller microcosm of the human and animal body. He believed that the circulation of the blood was analogous to larger cycles in nature, including the motion of the heavens and the cycle of rain on earth. The circle is a recurring image in his work. To the Greeks, it represented perfection. The heart acted like the sun, which caused the cycle of rain. The sun's warmth caused rejuvenation and growth. Just as the heart and the sun acted as monarchs of their realms, Harvey believed too that God had given earthly kings the right to rule. The circulation of the blood and the cycle of rain also emulated the perfect, eternal circular motion of the heavenly bodies. By means of these analogies, the circulation displayed the meaning and purpose of the universe.

Harvey's significance for science lay not only in his discovery of the circulation. He revealed the immense power of experimental demonstration, especially animal experimentation, to disclose valuable information about the body. In his second major work, *Anatomical Exercises on the Generation of Animals* (1651), Harvey employed his experimental and observational skills to uncover the secrets of generation, including the relative roles of male and female, how fertilization occurs, and the early development of the embryo. He refuted the "erroneous and hasty conclusions" of the ancients; "like phantoms of darkness," he claimed, "they suddenly vanish before the light of anatomical inquiry."* Harvey investigated frogs, snakes, fish and other marine animals, insects, and a number of domestic fowl and mammals. Like many others, he traced the development of the embryo by examining the stages of development in chicken eggs. Harvey went on to argue that the egg was the unit of development for all animals. The title page of *On Generation* proclaimed, "ex ovo omnia" (all is from the egg).

As a physician to the English King Charles I, Harvey had access to almost unlimited numbers of deer for dissection. An avid hunter, the king shot deer at one of his many estates nearly every week. Harvey's work on these animals, coupled with vivisections of dogs, cats, and rabbits, allowed him to give a minute account of fertilization, gestation, and embryonic development. From his observations of one stage of development Harvey devised questions and hypotheses that he then attempted to demonstrate on animals. This close observation of large numbers of animals was as important to subsequent research as were the experimental techniques of *De motu cordis*.

Harvey displays none of the showman's glee of Vesalius in his descriptions of vivisection. Animals, to Harvey, served a purely instrumental function, and if the question of cruelty occurred to him, he never expressed it. A modest and appealing persona emerges in his books, but Harvey did not hesitate to dissect the body of his own father (in front of an audience) in his quest for knowledge. Such an act was more shocking to contemporaries than the vivisection of animals. In this era, public dissection and vivisection flourished alongside many forms of public entertainment based on animal performance. Dancing bears, bears fighting with dogs, cockfighting, and all manner of cat torture were commonplace, and everyday cruelty to animals was the rule rather than the exception. Petkeeping was still uncommon, although Harvey described his wife's pet parrot, whose death she grieved. Her feeling for the bird, however, did not prevent her spouse from dissecting it.

Like Aristotle, Harvey was a vitalist who believed that living creatures differed fundamentally from non-living things. That difference could not be analyzed, or seized by experimental art. The ultimate cause of life was unknowable, and the best science merely described its effects. To Harvey, the heart was simply an intermediate cause, not the secret of life, and the ultimate cause of generation could not be found by humans. The blood was for him the critical element; it was the agent of vital heat, the very essence of life. The blood is the

---

*William Harvey, *Anatomical Exercises on the Generation of Animals* (1651), trans. Robert Willis, 1847 (New York: Johnson Reprint, 1965), 151, 164.

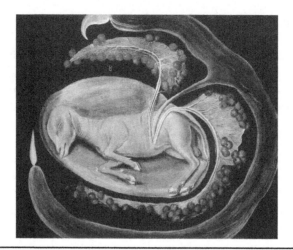

The Course of Gestation

In his work on generation, William Harvey describes the course of gestation in deer:

By the great number of dissections which I performed in the course of this month, I was every day confirmed in my opinion that the carunculae [fleshy protuberances] of the uterus perform the office of the placenta; they are at this time found of a reddish colour, turgid, and of the size of walnuts. . . . About the end of December the foetus is a span long, and I have seen it moving lustily and kicking; opening and shutting its mouth; the heart, inclosed in the pericardium, when exposed, was found pulsating strongly and visibly; its ventricles, however, were still uniform, of equal amplitude of cavity and thickness of parietes; and each ending in a separate apex, they form together

a double-pointed cone. . . . The internal organs, all of which lately had become perfect, were now larger and more conspicuous. The skull was partly cartilaginous, partly osseous. The hooves were yellowish, flexible, and soft, resembling those of the adult animal softened in hot water. . . . I have remarked an admirable instance of the skill of nature, in the bulge or convexity of the caruncles turned toward the conception.

■ Quote from William Harvey, *Anatomical Exercises on the Generation of Animals* (1651), trans. Robert Willis (1847) (New York: Johnson Reprint, 1965), 494–96. Illustration: Fetal sheep, Hieronymus Fabricius ab Aquapendente, *De formato foetu* (Venice: Franciscus Bolsetta, 1600). Courtesy of the Library of the College of Physicians of Philadelphia.

first to live and the last to die, and Harvey called it "the immediate instrument of the soul," the bearer of the vital principle endowed by God. The purpose of circulation was to revivify and reheat the blood that cooled off as it moved through the body.

Harvey sought to answer the old Aristotelian question of purpose. He concluded that life is autonomous and spontaneous and cannot be reduced to mechanics, to the mere analysis of matter and motion. But his image of the heart as a pump was such a compelling mechanical image that Harvey's interpretation, resisted by many of his Galenist contemporaries, was seized upon most of all by the emerging mechanical philosophers, who believed that all phenomena were reducible to mechanics and saw no need for irreducible, mysterious vital functions.

Descartes, the Villain of History

Despite Harvey's sacrifice of hundreds of animals, he has been ignored by modern antivivisectionists and animal rights activists. The role of the villain in the story of the use of animals in science has instead been assigned to Harvey's younger contemporary, the French natural philosopher René Descartes (1596–1650). In 1982, a member of the radical Animal Liberation Front slashed a portrait of Descartes at the Royal Society in London. The Australian philosopher Peter Singer, author of *Animal Liberation* (1975), calls Descartes's ideas about animals the "absolute nadir" of Western thought on that topic. In *The Case for Animal Rights* (1983), American philosopher Tom Regan spends an entire chapter attacking Descartes's views; "it is tempting," he admits, "to dismiss Descartes's position . . . as the product of a madman."*

All this obloquy is directed at a man who performed very few experiments on animals, who liked his pet dog and, as far as we know, did not dissect him. Descartes's villainous reputation stems from his philosophical claim that animals are not conscious. Therefore, he believed, they lack the ability to suffer and feel pain in the way humans do. A modern medical ethicist repeats a widely believed view that "Descartes's denial that animals (despite all appearances to the contrary) were able to suffer, appears to have been widely used as a justification for experimenting on live animals, at a time when that practice was becoming more common."† But most of his contemporaries did not accept Descartes's views, which are not as straightforward as we have been led to believe.

Unlike Harvey, who remained in many ways an Aristotelian, Descartes was the chief theorist for a new philosophy of nature known as the mechanical philosophy. The central premise of the mechanical philosophy as it developed during the seventeenth century was that the universe was a machine, which

---

*Peter Singer, *Animal Liberation,* 2d ed. (1990; reprint, New York: Avon, 1991), 200; Tom Regan, *The Case for Animal Rights* (Berkeley: University of California Press, 1983), 5.

†Institute of Medical Ethics, *Lives in the Balance,* ed. J. A. Smith and K. M. Boyd (Oxford: Oxford University Press, 1991), 300.

operated according to the laws of mechanics. In 1543, the same year Vesalius published *De fabrica*, the Polish astronomer and priest Nicolaus Copernicus (1473–1543) published his *On the Revolutions of the Heavenly Spheres (De revolutionibus orbium coelestium)*, in which he proposed that the sun and not the earth was the center of the universe, and that the earth was therefore one planet among many. The implications of this idea were only gradually realized. By the end of the sixteenth century, Galileo Galilei (1564–1642) and others recognized that if the earth was no longer considered the center of the universe, then all of physics required rewriting.

While Galileo focused on the problems arising from Copernicus's astronomy, Descartes intended to systematize Galileo's mechanics and his mathematical approach into a new philosophy. This mechanical philosophy of nature separated the psychic and spiritual world from the physical and material world and established the primacy of material (rather than other sorts of) causation. Matter itself was not active, but inert or dead. Theologians, such as Descartes's Jesuit teachers, feared that a clockwork universe governed by the laws of motion could operate without God's intervention. Descartes developed a philosophy of nature that would, he thought, incorporate the new mechanical world-view while keeping intact God's central role in the governance of nature. Galileo's works were condemned by the Catholic Church, and he was placed under house arrest in the 1630s. Descartes hoped to avoid this fate.

While traveling as a soldier in the winter of 1619, Descartes spent a rare day alone in thought and developed the basis for his new philosophy. He later described this remarkable day's thoughts in his *Discourse on Method (Discours de la méthode*, 1637). Descartes attempted to empty his mind of all preconceived ideas and to find what he knew to be true instinctively or intuitively— what appeared to him to be "clear and distinct" ideas. He discovered his thinking mind and therefore himself, who possessed the mind: "I think, therefore I am" (in Latin *cogito ergo sum*). Descartes existed primarily, then, as a thinking being; his body was all but irrelevant. His second revelation was the idea of God who made him, the idea of whom was so clear and distinct that it must be true. Descartes's third principle was the truth of geometry and mathematical relationships, which were self-evident *whether or not they referred to actual objects*. These self-evident principles were axiomatic and therefore not capable of analysis.

From this basis Descartes set out to reconstruct the world. He viewed the world as a collection of mechanisms that could be looked at and analyzed only in mathematical and mechanical terms. By this argument, there was no distinction between what we would call the physical and biological worlds, no special vital force. Everything could be analyzed in terms of the laws of mechanics. But reality did not necessarily correspond to our perception of it. To explain visible appearances in terms of matter in motion, he and other mechanical philosophers turned to the ancient doctrine of atomism, which stated that visible matter was composed of small particles individually invisible to the eye. The evidence of the senses, therefore, did not tell the whole story and could be positively misleading. Color and texture were merely the result of the particular disposition of the minute parts of matter.

Descartes proceeded, therefore, not by experimentation but by reason, and his discussion of the motion of the heart shows his fundamental differences from Harvey. Descartes learned of Harvey's theory of the circulation soon after its publication, and he probably witnessed a public demonstration of this theory in the Netherlands in the early 1630s. In his *Discourse on Method* he attempted to restate Harvey's discovery in mechanical terms. While Harvey had refused to assign a cause to the heart's motion, Descartes did not hesitate to assign a mechanical cause. While Harvey's analogy of the heart to a pump was simply that—an analogy, not a description—Descartes was convinced that the mechanisms he described were real.

Descartes did not describe the process of dissection or vivisection, but simply the result, for he would not demonstrate the circulation through dissection, but by means of mechanical principles, by logic. The question of animal heat remained critical, but Descartes transferred its site from the blood to the heart, and the mechanical effects of the heat of the heart on the blood gave rise to the circulation. He did not measure that heat, and no numbers appear in his account. The heart's heat derived from fermentation, like that of wine, a mechanical process caused by the motion of small particles. It was not the result of a mystical vital force as Harvey contended. The heartbeat occurred when the heart's heat caused the particles of blood to expand and turn into vapor. The vaporized particles of blood then spewed into the arteries, causing the heart to contract and draw in more blood. The image is that of a boiling teakettle (constantly refilled with cold water) rather than a pump. By this account, the dilating (diastolic) motion of the heart was the active phase of the heartbeat, the opposite of Harvey's view. Descartes claimed nonetheless that his description was consistent with observed phenomena. The configuration of the parts, he said, dictated this consequence as much as the wheels and weights of a clock determined its motion.

The human body was, therefore, a machine. But humans also possessed reason. Descartes's first clear and distinct notion of himself was that he was thinking, not that he had a body. Since humans could think, they necessarily had knowledge of God (Descartes's second clear and distinct idea), and therefore they possessed immortal souls. The essence of the world, to Descartes, was this dualism of mind and body, the complete separation of matter and spirit—a theological as well as a philosophical principle.

As Thomas Aquinas had been impressed with the new mechanical clocks of the thirteenth century, so Descartes marveled at the clockwork automata of his time, such as the famous mechanical fountain at St. Germain-en-Laye outside Paris. If humans could make such convincing devices, how much more clever was the hand of God, who made the infinitely more complex animal machine. In the *Discourse on Method,* Descartes said that humans could conceivably make a mechanical animal that would be indistinguishable from the real thing. But a mechanical human could never be mistaken for a real human, because it would lack a mind, which only God could bestow and which God only bestowed on humans. The mechanical human would manifest its inadequacy in two critical respects: it would lack speech, and it would lack the ability to reason.

Descartes believed that animals could neither speak nor reason, and therefore they were simply body, mere machines. The arrangement of their parts could lead animals to emit sounds in response to certain stimuli, or to act in certain ways. God had constructed the body, said Descartes, to do quite a lot without reference to the mind. While he argued that animals were not self-conscious or capable of reasoned thought, they did, he believed, feel pain, heat, hunger, and emotions such as fear and joy. The mind's functions included memory, conscious perception, and most important, reason. As a Catholic who constantly sought the approbation of the Church, Descartes knew that Church doctrine did not grant animals an immortal soul. He had observed Galileo's fate in the early 1630s and he so feared the wrath of the Church that he suppressed his major statement of a mechanical universe, *Le monde (The World,* 1633, published 1664). On the existence of animal intelligence and soul, Descartes sided with the Church against Michel de Montaigne (1533–1592), a skeptical philosopher who doubted the possibility of attaining true knowledge, and his disciple Pierre Charron (1541–1603). Montaigne argued that the complexity of spiders' webs and beehives implied that their makers possessed intelligence, even superior intelligence to humans. In his "Apology to Raimond Sebond" (1580), Montaigne satirically claimed that animals surpassed humans in industry and in cruelty. Charron took this view more seriously, and in his 1601 book *Traité de la sagesse,* argued that animals have souls.

Descartes dissected dead animals, but he seldom experimented on live animals. Unlike Harvey, Descartes did not regard experimentation as a method of discovery but as a way to confirm what he had already deduced by mechanical principles. By deduction, Descartes could generate several plausible mechanical scenarios to explain a phenomenon. Experiments could help to decide which of these scenarios was correct. Many of Descartes's experiments served simply to extend observation, to display the true arrangement of the parts. While Harvey believed that the soul was contained within the body, Descartes (hewing far more closely to theological orthodoxy) believed that the soul was essentially different from the body, and that learning about the body—especially the animal body—could tell us little about the soul.

"The Martyrs of the Anatomists": Animal Research in the Seventeenth Century

Descartes accepted Harvey's animal experimentation as a valid method of research, but conservative critics of Harvey argued that animal experimentation provided invalid evidence. They contended that experiments on animals could tell us nothing about humans because humans were unique. Some also revived the old empiricist argument cited by Celsus, that vivisection caused pathological changes in the research animal that invalidated the data. However, most researchers eagerly followed Harvey's methods, although most rejected his vitalist philosophy.

Almost all researchers in the second half of the seventeenth century began their research with two fundamental assumptions: that organisms and machines are analogous, and that all vital functions involve only the substances and processes of inanimate nature. There is no distinction between physics and physiology, no special vital force in nature. The Cartesian notion that animals were machines (the "beast-machine" concept as it came to be dubbed) was far more important as a methodological principle, an approach to research, than as a moral principle or a description of what animals were really like. Most researchers looked for evidence of mechanism in animal form and function, but they did not necessarily believe that the animals upon which they experimented felt no pain. Two groups of researchers were especially important: the circle around Robert Boyle at the Royal Society in London, and Marcello Malpighi and his circle in Italy.

As the King of England's physician, William Harvey followed Charles I from London to Oxford in 1642 after civil war broke out. He remained there until the town fell to Parliamentary forces in 1646. In Oxford, Harvey found a congenial circle of men with whom he discussed his work and performed experiments. This group continued to work in Oxford after Harvey left, and in 1660, when Charles I's son was restored to the English throne, they began a scientific club in London called the Royal Society where they continued their experimenting. Their research revolved around problems Harvey had introduced with his discovery of circulation. These included the purpose of respiration, the cause of animal heat, how metabolism works, why arterial blood is red and venous blood dark, and the cause of nervous transmission. However, they sought mechanical causes and not vital principles.

The Royal Society circle used many animals, alive and dead. Their techniques included the ligature of blood vessels, the inflation of lungs or other organs to show their structure, and the injection of various substances. Among the best-known experiments in this period were those employing a vacuum pump, or air pump, invented in the late 1650s by Robert Hooke (1635–1703) and Robert Boyle (1627–1691). Boyle and Hooke used the air pump to determine whether animals needed air to survive. Mice, cats, birds, snakes, and cheese-mites were put in the vacuum and also simply left inside the bell jar of the pump without withdrawing air. Boyle and Hooke concluded not only that animals needed air, but that something in fresh air as opposed to old (respired or burned) air was essential to life. Air was a dynamic and necessary element in sustaining life.

Hooke and Richard Lower (1631–1691) undertook a series of surgical vivisections in the mid-1660s to determine the mechanism of respiration. In these experiments, the thorax and diaphragm of a dog were cut away, leaving its heart visible. The dog was kept breathing by pumping air into its windpipe through a bellows. As soon as the bellows stopped, the dog's heartbeat became irregular. While this experiment showed that air was essential to life, it could not disprove the claim that the motion of the lungs (as some scientists contended) was also essential. Hooke and Lower repeated this experiment a few years later, with a twist. They used two sets of bellows, which kept the lungs continuously full but motionless (they pricked the pleural membrane so air escaped out the bottom). The heart continued to beat even when the lungs did not move, proving that the air, not the motion of the lungs, was the critical factor.

Some of the spectators at the Royal Society, while impressed with the results, expressed unease at this research program. John Evelyn, a fellow of the Royal Society, described the open thorax experiment as "of more cruelty than pleased me."* He also disliked dog fighting, a popular spectator sport. Like many of his contemporaries, and few of his ancestors, he owned several dogs as pets. Hooke himself disliked the open thorax experiment because of its

---

*E. S. de Beer, ed., *The Diary of John Evelyn*, 6 vols. (Oxford: Clarendon Press, 1955), 3: 497–98.

Pneumatick Engine

In 1659, Robert Boyle and Robert Hooke placed a lark in the receiver of their "pneumatick engine" and pumped out the air. The lark died, the first of many animals sacrificed to the air pump. Boyle's books and his papers in the *Philosophical Transactions* of the Royal Society, the first scientific journal, made the air pump widely known. This illustration of the second version of the air pump probably shows what Boyle called a "kitling" expiring in the vacuum. Hooke became curator of experiments at the Royal Society in the 1660s, and the air pump was a popular demonstration instrument at its meetings. It also proved to be an important step in the professionalization of science. The air pump was expensive and difficult to make, and it required special expertise to operate. Experiments that employed it could not be repeated by just anyone. This use of a machine may also have made it easier for researchers to think of the animals they used as mere machines as well.

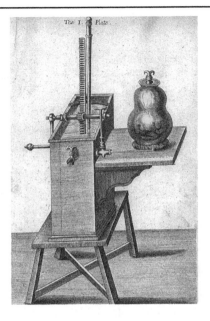

■ Robert Boyle, *A Continuation of New Experiments, Physico-mechanical, Touching the Spring and Weight of the Air, and Their Effects* (Oxford, Printed by H. Hall, for R. Davis, 1669–82). Courtesy of William Andrews Clark Library, University of California, Los Angeles.

cruelty, and it took three years for Lower to persuade him to repeat it. In his book *Micrographia* (1665), Hooke expressed his preference for the nonviolent observation of nature that the microscope afforded.

But Hooke continued to seek experimental demonstrations that could edify and entertain the Royal Society's membership of learned men and fashionable dilettantes. The technique of injection fulfilled the Society's requirements of dramatic effects with serious intent. Researchers tried to trace the course of blood vessels in animals with injections of ink. Others injected various drugs, poisons, and paralytic agents to show their mode of action. Attempts to feed intravenously with injections of broth and milk failed, but dogs got drunk when injected with wine and beer. The balance of the dramatic with the serious was not always equal.

Entertaining, horrifying, and irresistible (to judge by the amount of pub-

INJECTING BLOOD ROYAL or PHLEBOTOMY at S'CLOUD

Blood Royal

In his 1676 play *The Virtuoso,* Thomas Shadwell satirized the Royal Society's best-known experiments. The illustration, a caricature of Napoleon receiving a transfusion of "blood royal" from a tiger, shows that the idea of blood transfusion remained in the public consciousness long after its brief career in the seventeenth century. In this passage from Shadwell's play, the virtuoso, Sir Nicholas Gimcrack, explains his experiments to the "gentlemen of wit and taste," Longvil and Bruce and his friend Sir Formal Trifle:

LONGVIL: Sir, I beseech you, what new curiosities have you found out in physic?

. . .

SIR NICHOLAS: Why I made, sir, both the animals to be emittent and recipient at the same time. After I had made ligatures as hard as I could (for fear of strangling the animals) to render the jugular veins turgid, I open'd the carotid arteries and jugular veins of both at one time, and so caus'd them to change blood one with another.

SIR FORMAL: Indeed that which ensu'd upon the operation was miraculous, for the mangy spaniel became sound and the bulldog mangy.

SIR NICHOLAS: Not only so, but the spaniel became a bulldog and the bulldog a spaniel.

BRUCE: 'Tis an experiment you'll deserve a statue for.

■ Quote from Thomas Shadwell, *The Virtuoso* (1676), ed. Marjorie Hope Nicholson and David Stuart Rodes (Lincoln: University of Nebraska Press, 1966), 47–48. Illustration: "Injecting blood royal or phlebotomy at St Cloud," 1804. Courtesy of National Library of Medicine.

licity they received), the blood transfusion experiments of the late 1660s dramatically displayed the strengths and weaknesses of the experimenters' program. The transfusion craze swept Europe, especially England and France. The composition of the blood was imperfectly understood, and the concept of blood types unknown. Many physicians believed that transfusion had great therapeutic potential. Even if experimenters did not agree with Harvey that the blood contained the soul, older ideas lingered of its mystical character and role as the repository of temperament. A German doctor suggested that mutual transfusions between a husband and wife could lead to conjugal harmony. The central ritual of Christianity transformed wine into the blood of the Savior. Furthermore, despite Harvey's convincing proofs that the body could not generate an infinite supply of new blood, bleeding continued to be a common therapy.

Animal-to-animal transfusions ran the gamut, as blood was exchanged between "Old and Young, Sick and Healthy, Hot and Cold, Fierce and Fearful, Tame and Wild Animals."* At the Royal Society, Richard Lower and his colleagues began with dog-to-dog transfusions but soon tried cross-species transfusions with sheep and calves as well as dogs. Perhaps it was inevitable that it would be tried on humans as well. The experimenters argued that the blood of young calves and lambs was purer than human blood, less tainted by human passions and vices. Did not the blood of the lamb symbolize the pure healing blood of Christ? Healing the passions of the mind was an especially important goal of these early transfusers. Yet their overly passionate (if not quite insane) subjects were also in no position to complain about their treatment. In these early attempts at human experimentation, no notion of "informed consent" arose, and neither the patients nor the transfusers seriously questioned the morality of the experiment.

A French royal physician, Jean-Baptiste Denis, transfused the blood of a sheep into a young man suffering from a fever, whose own blood supply had been depleted by therapeutic bleeding. Denis reported in the *Philosophical Transactions* that the young man appeared recovered. Lower and a colleague soon tried a similar experiment at the Royal Society, transfusing sheep's blood into a young clergyman whose brain had been described as "sometimes a little too warm." The transfusion appeared to have the desired effect of calming the young man's behavior.† A second transfusion on the same individual a few

---

* Richard Lower, "The Method Observed in Transfusing the Blood out of One Animal into Another," *Philosophical Transactions* 2 (1666): 353–58.

† Edmund King to Robert Boyle, 25 November 1667, in Robert Boyle, *Works*, ed. T. Birch (London, 1772), 6: 646–47.

weeks later brought out a crowd of spectators. Shortly thereafter Denis transfused a similarly manic patient, again reporting success. However, the patient died following an attempt at a third transfusion a few months later. Both the French and English patients complained of various symptoms after the second transfusion. They were experiencing a *hemolytic reaction* to the foreign blood, characterized by a massive die-off of their own red blood cells. That they did not experience reactions sooner can probably be attributed to the inefficiency of the transfusion process; it is unlikely that they received more than a few ounces of animal blood. However, that would be enough to stimulate their bodies to form antibodies to the foreign blood, and subsequent transfusions would lead to more severe consequences. The death of Denis's patient early in 1668 quickly ended any further attempts at animal-to-human blood transfusion. Experimental blood transfusion only resumed in the 1820s, and human-to-human blood transfusion did not become an accepted medical practice until after the discovery of blood types in the early twentieth century.

Although the transfusion experiments attracted much public attention, they were only a small part of the research enterprise in this period, which was a fiercely contested territory of competing theories. Most researchers agreed that Harvey's theory of the circulation was true and even agreed that his anatomical methods were useful, but the interpretation of his discovery, as we have already seen in the case of Descartes, could vary widely. Boyle and his circle were mechanists in the sense that they believed the animal body could be analyzed in mechanical terms. But they did not follow Descartes in attempting to discover or hypothesize the micromechanisms that made the body function. While national rivalries, especially between the French and English, played a role in the transfusion experiments, new scientific journals also ensured that competing researchers shared their ideas with one another and with the interested public. Public dissections were another forum for researchers to display new ideas.

The Italian anatomist Marcello Malpighi (1628–1694) and his colleagues also experimented on animals, but their experiments were very different from those performed at the Royal Society. Rather than physiological processes such as respiration, Malpighi investigated the microstructure of the lungs and other organs. Over a period of years, he and his colleague Carlo Fracassati (d. 1682) dissected and vivisected guinea pigs, cats, sheep, frogs, and birds. Over and over again, Malpighi injected the lungs and the pulmonary vessels with colored water to determine their structure. The lungs were revealed as a series of small cellular vesicles separated by membranes, rather than solid or even spongy flesh. More difficult was the tracing of the pulmonary circulation. No one had yet seen the anastomoses, or connections, between the smallest veins and capillaries, even though Harvey's theory required them. Malpighi's injection methods encountered technical problems: in dead animals, the fine vessels were clogged with clotted blood, while in live animals, the blood flowing through the vessels could not easily be replaced with colored water without killing the animal.

Malpighi's correspondence with his friend Giovanni Alfonso Borelli (1608–1679) over a period of two years reveals the hit-and-miss nature of experimentation: one method failed, he tried another, Fracassati reported different results, Borelli made suggestions, Malpighi tried again, a different animal, a different technique. The anastomoses must exist. But how to prove it? Fracassati and then Malpighi tried frogs. Borelli complained that he did not have access to frogs and could not duplicate Malpighi's observations. Malpighi nonetheless continued, sacrificing, he said, almost the entire race of frogs.*

Frogs were excellent subjects for this research: their lungs are relatively simple in structure, and nearly transparent, allowing a clear view of the blood vessels within. Malpighi observed the living frog, and then tied off the lungs top and bottom, retaining the blood within. He dried the tiny lungs, inflated them, and then examined them with a microscope. There he found the connections he sought. His remarkable treatise on the lungs, *Epistolae de pulmonibus,* published in 1661, gave new prominence to the newly invented microscope as a research tool. The microscope revealed the minute subtlety of nature, which atomism and mechanical theories only surmised. Malpighi hypothesized that the lungs functioned to break up the blood into ever more minute particles that would then circulate through the body to become bone, organ, or fluid. Nutrition was not mysterious, but a purely mechanical process. He rejected aspects of Cartesianism as too speculative, yet it was clear to him that the animal was indeed a machine.

By the end of the seventeenth century, the animal body (and by analogy, the human) was much better known than it had been at any time in the past, and scientists pointed to their own successes to justify their experimental practices against critics. Robert Boyle defended the practice of natural philosophy, including animal experimentation, in his *Some Considerations Touching the Usefulness of Experimental Natural Philosophy* (1663). A devout Christian, Boyle refuted claims that science threatened religion, referring to the "two book"

---

*H. B. Adelmann, ed., *The Correspondence of Marcello Malpighi,* 5 vols. (Ithaca: Cornell University Press, 1975), from vol. 1.

Beast-Machine

The Swiss physician Albrecht von Haller (1708–1778) experimented on dogs in his laboratory, attempting to demonstrate the sensations created by various forms of stimuli, including painful ones. Clearly Haller believed that these dogs felt pain. Only the most fervent Cartesians embraced the "beast-machine" concept so far as to act out its consequences. In the seventeenth century, this included the religious order of Jansenists, centered at the Parisian monastery of Port-Royal. A secretary to the Jansenist fathers described their cruelty to animals in a much-quoted passage:

They administered beatings to dogs with perfect indifference, and made fun of those who pitied the creatures as if they had felt pain. They said the animals were clocks; that the cries they emitted when struck, were only the noise of a little spring which had been touched, but that the whole body was without feeling. They nailed poor animals up on boards by their four paws to vivisect them and see the circulation of the blood which was a great subject of conversation.

Another Cartesian clergyman and philosopher, Nicolas Malebranche (1638–1715), reportedly kicked a dog at his feet and responded coldly to a protesting observer, "So what? Don't you know that it has no feeling at all?" To Malebranche, the denial of animal suffering was also a point of theology. Human suffering, including that

of children, could be attributed to the original sin of Adam, which condemned all of humanity. Since animals were not descended from Adam, they could not suffer, for God would not have created pointless suffering.

■ Both quoted in Leonora Cohen Rosenfield, *From Beast-Machine to Man-Machine* (1940; reprint, New York: Octagon, 1968), 54, 70. Illustration: Albrecht von Haller, *Memoires sur la nature sensible et irritable, des parties du corps animal* (Lausanne: Marc-Mic. Bousquet, 1756–60). Courtesy of Biomedical Library, University of California, Los Angeles.

notion that contemplation of nature leads to a finer appreciation of God through his works. Boyle also revived the old concept of stewardship. God had entrusted the earth to humans to improve it by their activities, not merely to contemplate it. To Boyle, animal experimentation provided an excellent example of God's sanction of natural philosophy. Not only did it afford knowledge of God's creation to the experimenter, it also provided useful knowledge and underscored the purposefulness of the creation. To Boyle, it was obvious that God had provided animals to the anatomist to make experiments he could not pursue on humans.

Boyle and his colleagues were aware that the animals could suffer. Although most researchers were committed to some version of the mechanical philosophy, and many were convinced Cartesians, few of them took literally Descartes's argument that if animals were machines, they did not feel pain as humans did. Boyle noted the animals' distress in the course of an experiment, and a viper was "furiously tortured" under the influence of the vacuum.* Lower bled animals to the point of death, determining this point from the animals' struggles. While a Cartesian might argue that these struggles were purely automatic, Lower did not argue this; instead he used language that denoted pain and suffering. Carlo Fracassati injected a dog with vitriol (hydrochloric acid) and noted, "The Animal complain'd a great while . . . and observing the beating of his breast, one might easily judge, the Dog suffered much."† All of them believed that such suffering was preferable to human suffering for the sake of advancing knowledge.

Most seventeenth-century scientists did not believe that the "beast-machine" notion of Descartes implied that animals felt no pain. Few used the supposed insensitivity of animals as an excuse for experimenting on them. Another clergyman-scientist commented to a friend, "I wish [the Cartesians] could convince me as thoroughly as they are themselves convinced of the fact that animals have no souls!!"‡ Rather, the concept of the "beast-machine" described a highly successful approach to research. Many scientists and observers in this period nonetheless began to experience pricks of conscience at animal experimentation.

* Robert Boyle, "New Pneumatical Experiments about Respiration," *Philosophical Transactions* 5 (1670): 2044.

† Carlo Fracassati, "An Account of Some Experiments of Injecting Liquors into the Veins of Animals," *Philosophical Transactions* 2 (1667): 490.

‡ Niels Stensen to Thomas Bartholin, 1661, quoted in Frederik Ruysch, *Dilucidatio valvularum in vasis lymphaticis et lacteis* (1665), ed. A. M. Luyendijk-Elshout (Nieuwkoop: B. de Graaf, 1964), Introduction, 36.

Beginnings of Moral Concern

Galen wasted no thought on moral considerations when he cut open living animals, and in this he was a man of his time. By the time Vesalius duplicated some of Galen's experiments in the 1530s, social standards had changed. In the Renaissance the humanist ideal of a cultured individual was superseding the warrior ideal of the medieval knight. The path to success in the sixteenth century was as much at the royal court as on the battlefield, and skills in polite conversation were valued. The ideal humanist was also a Christian who would display kindness and compassion.

Vesalius seems little different from Galen in his feeling for animals. But the spectacle of public anatomy, which included demonstrations on animals, was intended (unlike in ancient Rome) to have a moral impact on its audience. Viewing the end of all flesh reminded the audience of the importance of their souls, and the suffering of the animals induced compassion. In the finale of his public anatomy, Vesalius vivisected a pregnant animal, either a sow or a dog, expertly manipulating both the animal and his audience.

Vesalius's successor as professor of anatomy at Padua, Realdo Colombo, cut open a pregnant dog, removed the puppies, and then hurt them in front of the mother. Ignoring her own pain, she tried to comfort the pups. Colombo reported that the bishops and other clergymen in attendance were especially impressed by this display of motherly love. The juxtaposition of life and death, pain and pleasure, conveyed a powerful emotional impact, and the fact that the sufferer was animal rather than human did not diminish this impact. In public anatomy, animals acted as moral and physical proxies for humans. Thomas Aquinas had declared that being cruel to animals could lead to being cruel to humans. But Thomas did not condemn the suffering of the animal as morally objectionable in itself.

In the seventeenth century, as the use of animals in experiments increased, popular attitudes toward animals also began gradually to change. Both public anatomy and the use of animals in popular entertainments such as bear-baiting and public combats continued well into the eighteenth century. But public anatomy disappeared around 1800, and by that time animal sports (with the notable exception of hunting) were increasingly viewed as lower-class activities, too disturbing for educated tastes. Meanwhile petkeeping increased: English monarch Charles II's mistresses had their portraits painted with their lapdogs, and a popular English pamphlet of 1644 bemoaned the death of a pet dog in battle.

Public knowledge of scientific practices meant that they were not immune from public criticism. In France, the notion of the "beast-machine" came in for strong criticism. Most scientists believed that animals could feel pain, and several Jesuit intellectuals expounded on this theme. In his *Discourse on Animal Knowledge* (1672), Father Ignace-Gaston Pardies relied on Aristotle to support his argument that animals can feel, imagine, and remember, although they cannot contemplate or have spiritual knowledge. Moreover, Pardies continued, God would not have given animals sense organs just for show. In 1690, another Jesuit, Gabriel Daniel, extended Pardies's arguments in his *Voyage to the World of Descartes*. Daniel cited animal experimentation as an instance of extreme cruelty, arguing, like Thomas Aquinas, that cruelty to animals would lead to cruelty to humans. Daniel did not, however, argue that animals have rights, a notion that would have seemed absurd to him.

John Ray (1627–1705), an English clergyman and naturalist, attacked the "beast-machine" notion on theological and on scientific grounds. In a 1693 essay, Ray took the theological position that God did not create the world for humans alone. Animals existed of themselves, not merely for our use, and expressed God's creative power. He argued that animals were conscious but not rational. This argument did not put animals under moral consideration, but Ray nonetheless believed that animal suffering was immoral, declaring, "the torture of animals is no part of philosophy."* The debate had begun.

Harvey's discovery of the circulation brought animal experimentation to the forefront as a scientific method. The mechanical philosophy of nature, most forcibly enunciated by Descartes, gave an added dimension to this method as researchers sought mechanical explanations for the operations of the human and animal body. While some followed Descartes in seeking actual micro-mechanisms, others viewed mechanism as a convenient metaphor. Few believed Descartes's contention that animals' lack of consciousness meant that they could not in some way feel pain. The increased number and invasiveness of experiments led some to consider animal suffering as a moral issue for the first time since antiquity.

---

* John Ray, "De animalibus in genere," in *Synopsis methodica animalium quadrupedum et serpentini generis* (London, 1693), 12, translated in Charles Raven, *John Ray Naturalist* (Cambridge: Cambridge University Press, 1942), 375.

3 Disrupting God's Plan

Shortly before Christmas in 1694, Queen Mary II of England, aged thirty-two, died of smallpox. Her death set in motion a chain of events with profound dynastic consequences for the English crown. Mary and her Dutch husband William of Orange had assumed the English throne six years earlier when her father, King James II, was deposed. Mary and William had no children, and her death, in the middle of her childbearing years, meant that the heir to the throne was now her sister Anne, whose only son, the Duke of Gloucester, was not at all healthy. When the young Duke died of smallpox in 1700, Parliament took the question of the succession into its own hands. The Act of Settlement in 1701 placed the line of succession in a set of German cousins rather than allowing James or his children from his second marriage to return to England. The first of the kings from Hanover, George I, took the throne in 1714. He spoke no English.

Thus could disease determine the fate of dynasties and nations. Although, in this age before antibiotics, many infectious diseases were endemic, smallpox was an especially omnipresent affliction. The virus that caused it could travel through the air, making it highly contagious, and simply being in the same room with a smallpox sufferer could convey the disease. The course of the disease included a high fever, upper respiratory symptoms, and finally the characteristic "pocks" or rash. Each stage of the disease carried its own dangers, but the pocks were the most dangerous and most feared. Discrete pocks were better than confluent ones that ran together into one suppurating sore, but in either case the pocks could attack the eyes as well as the inside of the mouth and throat. What we now recognize as different strains of the smallpox virus could cause these symptoms to be more or less severe. How smallpox reached Europe was uncertain, but by the sixteenth century it had become established as one of the more serious of the many hurdles facing children in their struggle to reach adulthood. If they reached adulthood without catching it, they were by no means safe from it. The pockmarked faces and bodies of many adults, and the blindness of others, gave evidence of that struggle. Many more did not survive the disease; in an average year, about 20 percent of all those infected died. In years of epidemic, a larger percentage perished. During the epidemic of 1721, the parish of St. Giles-in-the-Field in London recorded sixty-eight deaths during the week of March 14. Smallpox had claimed nineteen of those.

Prevailing medical theories offered conflicting advice to the smallpox victim. While the disease was obviously contagious, no one knew exactly how a sufferer conveyed it to others. Especially in its early stages, smallpox was difficult to distinguish from other rash-producing diseases such as measles. Surviving an attack was believed to confer immunity, but no one knew how or why this occurred. The French king Louis XV had been diagnosed with a mild case of smallpox in 1728, and was thus thought to be immune. When he became ill in 1774, smallpox was not suspected until far too late, and Louis died miserably. But what, in any case, could his doctors have done? Some physicians advocated keeping the patient warm, while others urged a cooling therapy to counteract the effects of fever. Amulets were used to ward off the disease, and noxious potions of tar water or sheep's dung were used to cure it. Patients were bled and purged, although doctors disagreed about the correct time for these procedures. Some still advocated the medieval "red therapy" of surrounding the patient with the color red to "draw out" the pocks and thus draw the disease out of the body. England's Queen Elizabeth I had supposedly been cured of an attack in 1562 by a combination of two therapies: she was wrapped in red flannel, and she was placed before the fire to keep warm. Her royal descendant Mary II was less fortunate.

Smallpox was one of the diseases Europeans had brought to the New World that had decimated indigenous populations. Many in Europe believed it was increasing in severity in the seventeenth and eighteenth centuries, and smallpox epidemics such as in London in 1681 seemed to confirm this belief. Among childhood diseases, it had a high rate of fatality (as much as 40 percent in some years), and the disfigurement of those who survived was a lifelong affliction. It was among the most loathed and feared, and most common, of diseases. These factors led a number of Europeans to take part in the largest human experiment ever conducted: the smallpox inoculations of the eighteenth century. The inoculation controversy raised several issues that would reappear in later discussions of human experimentation, particularly in twentieth-century clinical trials. These included the concept of acceptable risk, the amount of information the patient should have, and the physician's responsibility to the community.

It had long been recognized that those who had once had smallpox did not usually get it again. Some parents would allow their uninfected children to catch the disease from their siblings, and in rural areas, children were often urged to get in bed with relatives suffering from smallpox. In other areas, "buying the pox" was also common. Parents would send their children to the home of a victim of a mild attack to purchase a few of the scabs fallen from the pustules. African slaves reported similar practices in their homelands of deliberately inducing the disease to confer immunity. But no one understood how the disease was contracted, or how immunity was conferred.

Reports from outside Europe of deliberate inducement of smallpox to confer immunity began to appear at the Royal Society of London, the most famous scientific society in Europe, at the beginning of the eighteenth century. In 1700, a physician described to the Society the Chinese practice of inhaling dried smallpox matter to induce the disease. In 1713 and 1714, accounts appeared in the Society's *Philosophical Transactions* of the Turkish method of inoculation, in which matter from smallpox sores was inserted into small cuts in the skin, usually in the arm. This was similar to the African practice, and was widespread in areas of the Middle East.

European physicians were reluctant to take up the practice of inoculation even though it was apparently successful. Its association with folk tradition made it suspect, especially since it appeared to have connections with superstitious practices that had been discarded in Western medicine. In some areas, the inoculated arm was wrapped in a red cloth, a version of the "red therapy" of earlier centuries. In addition, physicians questioned whether a treatment that worked in a hot country would work in other climates and on other peoples. Despite long years of use elsewhere, the practice remained untested in Europe, and physicians were notoriously conservative in their therapies, as was evident in the continued use of bloodletting, even after Harvey's theory of the circulation would, it seems, have made the practice illogical. And what physician who valued his reputation would deliberately induce a disease? The general sentiment was that inoculation was experimental, in every sense of the term, and no physician wished to be the first to perform the experiment on his patients.

The two individuals who did the most to promote smallpox inoculation in the early 1720s were not physicians. Cotton Mather (1663–1728) in New England was a well-known Puritan clergyman as well as a regular reader of the *Philosophical Transactions,* where he first learned of inoculation. Lady Mary Wortley Montagu (1689–1762) had witnessed inoculation in Constantinople as the wife of the British ambassador there from 1716 to 1718, and her young son was inoculated by the embassy's surgeon, a Scot named Charles Maitland, in 1718.

Lady Mary was an outspoken and forthright woman who was vocal in her dislike and mistrust of physicians. When an epidemic of smallpox attacked London in 1721, she called on Maitland to inoculate her three-year-old daughter, also named Mary. Maitland nervously complied, even though Lady Mary had refused to allow any physicians to be present as witnesses. This was the first recorded inoculation in England. Several physicians examined young Mary after the inoculation. She came down with a mild case of smallpox and survived. One of the physicians was so impressed he prevailed upon Maitland to inoculate his own six-year-old son.

Everyone concerned knew that this was literally a shot in the dark. Little Mary Montagu came through splendidly, but the two-year-old son of the Earl of Sunderland, inoculated a year later, came down with a severe case of smallpox and died. Yet smallpox engendered such fear and horror that parents made their own cost-benefit analysis and took the risk. As we shall see, later in the 1720s some physicians attempted statistical demonstrations of the efficacy of inoculation. But this was an age of risk taking; the wars of the seventeenth century had ended (only temporarily, as it turned out), but the new stock exchanges of London and Paris allowed a different kind of risk, and the South Sea Bubble of 1720 took investors on a dizzying ride from the heights of wealth to the depths of poverty after it crashed. A popular form of investment was the tontine, in which several contributed to a fund that would be turned over to the last survivor of the group.

Almost in spite of themselves, the physicians who examined young Mary Montagu were impressed. Sir Hans Sloane, physician to the King, wanted experimental proof, even though the very act of inoculation could be seen as an experiment, since no one could be certain of the outcome. But such an experiment could only be conducted on humans, because only humans could get smallpox. Like the Alexandrian physicians, Sloane turned to the most powerless people in society: prisoners. He asked the King for permission to inoculate several condemned prisoners from Newgate prison in London. The prisoners would volunteer their services in exchange for their release—if they did not die from the experiment. The King gave his permission. The British royal family, perhaps mindful of the past dynastic consequences of smallpox, was among the first to hear about Mary Montagu, and was quite interested in inoculation.

On 9 August 1721, Sloane recruited Charles Maitland to inoculate six pris-

The Turkish Practice of Inoculation

Lady Mary Wortley Montagu described the Turkish practice of inoculation in a letter to a friend. The engraving, which appears in an edition of her letters edited by her grandson, is from a contemporary portrait.

The Small Pox so fatal and so general amongst us is here entirely harmless by the invention of engrafting (which is the term they give it). There is a set of old Women who make it their business to perform the Operation. Every Autumn in the month of September, when the great Heat is abated, people send to one another to know if any of their family has a mind to have the small pox. They make partys for this purpose, and when they are met (commonly 15 or 16 together) the old Woman comes with a nutshell full of the matter of the best sort of small-pox and asks what veins you please to have open'd. She immediately rips open that you offer to her with a large needle (which gives you no more pain than a common scratch) and puts into the vein as much venom as can lye upon the head of her needle, and after binds up the little wound with a hollow bit of shell, and in this manner opens up 4 or 5 veins. . . . The children or young patients play together all the rest of the day and are in perfect health till the 8th. Then the fever begins to seize 'em and they keep in their beds 2 days, very seldom 3. They have rarely above 20 or 30 in their faces, which never mark, and in 8 days time they are as well as before their illness. . . . Every year thousands undergo this Operation, and the French Ambassador says pleasantly that they take the Small pox here by way of diversion as they take the Waters in other Countrys.

■ Quote from Lady Mary Wortley Montagu to Sarah Chiswell, 1 April 1717, in Lady Mary Wortley Montagu, *Selected Letters,* ed. Isobel Grundy (London: Penguin, 1997), 158–59. Illustration: "Lady Mary Wortley Montagu in Native dress," engraving, 1836.

oners, three male and three female, before a crowd of learned witnesses. The prisoners, aged between nineteen and thirty-six, trembled as they awaited the surgeon's lancet. Maitland duly performed the inoculations, but a few days later was dissatisfied with the appearance of the incisions and repeated the operation with fresh infectious material. Closely watched, five of the six came down with smallpox; one, it was discovered, had already had the disease. All of them recovered, and were duly released.

Then as now, August was a slow time for news, and the London newspapers seized upon the Newgate experiments with alacrity. Some reprinted the accounts from the *Philosophical Transactions*. Others expressed skepticism about the operation; *Applebee's Journal* noted that inoculation was "an Invention that had its Rise among the Populace, who were neither Men of Fortune, Character, nor Learning." Even after the apparently successful conclusion of the experiment, *Applebee's* remained doubtful: "any Person that expects to be hang'd may make Use of it."*

Several newspapers and pamphlets pointed out that Sloane had not proven that inoculation conferred immunity. In response, he sent one of the released prisoners, nineteen-year-old Elizabeth Harrison, to nurse a smallpox victim under Maitland's surveillance. Even though she had fulfilled her side of the experiment, her class and perhaps also her gender made her vulnerable to further exploitation. Illiterate and unemployed, Elizabeth Harrison's consent, from inside and outside of prison, was not informed and only marginally voluntary. Maitland reported that Elizabeth Harrison "lay in the same bed" with her patient "every other night"; when a young boy also came down with the disease, Maitland "obliged [Elizabeth] to ly every night with the boy; and to attend him constantly from the beginning of the Distemper to the very end; And thus, she continu'd for six weeks together without Intermission; or suffering the least Head-ach, or other Disorder; tho, indeed, she once had some heats and Little pimples; as Nurses commonly have under such Confinements."† Elizabeth Harrison's opinion of all this has not been recorded.

The continued good health of Elizabeth Harrison convinced a few more doctors that inoculation did indeed confer immunity, and they in turn began to inoculate their patients. In November 1721, the Princess of Wales announced that she would pay for the inoculation of all the orphans in St. James's Parish who had not yet had smallpox. This magnanimous gesture was actually another experiment to test the effects of inoculation on children. Although this grand plan fell through, Maitland did inoculate six more volunteers at royal expense who were then available to be examined by the public, and in the spring of 1722 five orphans from St. James's Parish underwent the experience and were similarly displayed. The Princess of Wales was convinced, and in April 1722 Maitland inoculated her two daughters, the Princesses Amelia and Caroline.

* Quoted in Genevieve Miller, *The Introduction of Smallpox Inoculation in England and France* (Philadelphia: University of Pennsylvania Press, 1957), 56.
† Maitland to Sloane, no date (but ca. 1721), British Library, Sloane MSS 4034, f. 18.

Royal validation led to a flurry of inoculation among the British aristocracy. Yet many questions remained unanswered: How did inoculation work? Why did some of those inoculated get mild cases of smallpox and others severe ones? How long did this induced immunity last? Couldn't inoculated victims spread the disease as readily as natural cases (they could, and did)? Could physicians justify causing disease, for whatever reason?

The activities of Maitland and Sloane took place within a narrow circle of people and on a small scale, and at first raised little opposition. Although Lady Mary Wortley Montagu had initiated the process, Sloane had quickly assumed leadership with the Newgate trials. Sloane was first physician to the king, president of the Royal College of Physicians of London, and the first physician to be named a baronet (a hereditary knighthood). He was the most powerful physician in London, and his advocacy of inoculation stilled many critics, at least for a time. At the same time in New England, however, inoculation was attempted on a much larger scale, and aroused furious debate that soon reached the home country.

In New England, the clergyman–natural philosopher Cotton Mather, fellow of the Royal Society, had learned of inoculation from his African slave and later read about it in the *Philosophical Transactions*. When a smallpox epidemic struck Boston in the spring of 1721, he encouraged the local physicians to try inoculation as a way of checking the disease. Unlike Lady Mary, who had Maitland at her side as well as her own son for corroboration, Mather had only his own knowledge and authority as a prominent clergyman. Boston's physicians, led by Dr. William Douglass, a graduate of Edinburgh and Leiden, were not convinced. Only Dr. Zabdiel Boylston (1679–1766) responded, inoculating his sons, several of his slaves, and a few patients. This soon led to a large public outcry against Mather, Boylston, and the practice of inoculation. Some argued that it only served to spread the disease by giving it to people who may not have acquired it naturally. The physician William Douglass condemned Mather and Boylston for their "rash and thoughtless Procedure in a *Medical Experiment* of Consequence," but held Mather especially responsible as an amateur usurping the physician's role.* In Puritan Boston, another argument was even more compelling: inoculation implied a lack of trust in God's overriding plan, and amounted to an attempt to supersede God's authority. It was, after all, a heathen invention, and by displaying a lack of trust in God's benevolent providence, its practitioners threatened to stir up God's wrath. Had not

---

* William Douglass, *Abuses and Scandals* (London, 1722), Introduction (n.p.).

that wrath already been adequately displayed in the epidemic itself? Many argued that Bostonians should be praying and begging for forgiveness rather than tempting God further.

A pamphlet war raged for several months, but Boylston continued to inoculate his patients, including Mather's son. While Boylston and Mather ultimately won the war of public opinion—only twelve died of the four hundred Boylston had inoculated, compared to five hundred dead of the thirty-six hundred natural cases—they played a dangerous game, ethically and medically. Since no one understood how inoculation worked, the severity of the induced cases was largely a matter of luck, and their critics were not entirely wrong in accusing them of spreading the disease. Inoculation did create new cases of smallpox, and those inoculated tended to act as if they were not in fact infectious, dismissing precautions against further spreading the disease.

In England pamphlets also began to fly. Reports of inoculation appeared in the *Philosophical Transactions* and were repeated, not always accurately, in the popular press. Clergymen declared that only God had the power to give disease and to take it away. Opposition also rose among medical men, uneasy at the notion of causing a disease and uncertain of the nature of infection. How did Sloane and his colleagues know that smallpox was indeed a single disease? As we have seen, disease theory, based on humors, focused on individual imbalance rather than infectious agents, and many physicians remained unconvinced that the disease induced by inoculation was "real" smallpox. Some also worried that the smallpox matter could be contaminated with other diseases—which was indeed often the case. The folk origins of inoculation also made it suspect; who would wish to trust the judgment of "a few *ignorant women*"?*

Douglass in Boston argued that inoculation, like blood transfusion, was a parlor game of natural philosophers and should not be attempted on the general population. Many of the natural philosophers in the Royal Society indeed supported inoculation, and they answered its critics with the authority of mathematics. While Mather, like the ancient empirics, argued that experience would be the most certain proof against dogmatic critics, physician-scientists such as James Jurin (1684–1750) employed numerical arguments. Lack of a convincing medical theory for the success of inoculation gave numerical arguments even more weight. Intellectuals in this age of Enlightenment firmly believed that to mathematize a problem was to explain it, and because num-

* William Wagstaffe, *A Letter to Dr Freind; Shewing the Danger and Uncertainty of Inoculating the Small Pox* (London, 1722), 5–6.

bers were believed to be innately true, a mathematical explanation was also true. Their model was the great English scientist Sir Isaac Newton (1642–1727), president of the Royal Society from 1703 to 1727, whose *Mathematical Principles of Natural Philosophy* (1687, known as the *Principia* from its Latin title) offered a mathematical explanation of the operation of the universe. Newton's sophisticated mathematics explained motion both in the heavens and on earth, fulfilling Galileo's aim of unifying the natural world by means of mechanics. Newton's scientific method, including his use of mathematics, his use of experiments (particularly in the *Opticks* [1704]), and his emphasis on effects rather than causes, was best illustrated in his discussion of universal gravitation. He demonstrated this concept by experiment, proved its existence mathematically, and refused to speculate on its cause. To demonstrate its existence was sufficient; unlike Aristotle, Newton did not seek ultimate causes, and indeed found them irrelevant to the pursuit of natural philosophy. His method and his discoveries were immensely influential in all areas of scientific endeavor in the eighteenth century.

James Jurin, a follower of Newton, was the secretary of the Royal Society in the 1720s. He used his wide network of correspondents to survey the effectiveness of inoculation by comparing the number of deaths caused by inoculation to those caused by naturally induced smallpox. Gathering case histories from a wide range of practitioners over a period of several years, Jurin collated this evidence and provided convincing proof that the mortality rate from natural smallpox far exceeded that from the inoculated version. Jurin analyzed the London Bills of Mortality (the official reports of deaths in the metropolis) for a twenty-year period to determine that the risk of dying from natural smallpox was about one in every eight cases; in times of epidemic, this could be one in five or six. In contrast, one in sixty had died among inoculated cases in New England, and Jurin arrived at a similar figure for the period of 1721 to 1727 in England.

Jurin's numerical arguments, based on the emerging science of probability, were quite convincing to contemporaries who sought answers in scientific authority. His arguments supported Newton's model of an orderly and balanced universe in which natural phenomena followed laws and were predictable. God, certainly, had designed this universe, not a wrathful God who punished sin with disease but a God who gave humans the tools to conquer disease. The age of miracles was over. Newton's ideas gradually spread across Europe, and numerical arguments eventually aided in the adoption of inoculation in other countries, particularly France, where physicians strongly resisted the idea of inoculation. Daniel Bernoulli (1700–1782), a mathematician and member of the French Academy of Sciences, wrote an analysis of smallpox mortality in 1760, arguing that out of thirteen thousand children, inoculation would save the lives of one thousand who would otherwise die of smallpox. By the end of the eighteenth century, inoculation was widespread in Europe, and itinerant inoculators, such as the Sutton family in England, inoculated thousands of people. Inoculation was risky, but taking one's chances on natural smallpox was far riskier. The mathematics of probability that Jurin and Bernoulli used was first developed in the seventeenth century to analyze games of chance, to predict who would win a bet. Inoculation was surely the biggest gamble of the eighteenth century, and it remained essentially an experiment, since no one knew how it worked, or why.

Mathematical analysis entered every science, including social science, in the eighteenth century, and inoculation was one of the first scientific uses of statistics, now so central to scientific practice. Jurin's case histories were boiled down to a stark numerical narrative. The individual patient became simply a number, all personal details effaced. This was an important step in the recognition of individual, identifiable diseases that acted the same in all victims, as opposed to the older view of a protean disease that differed in accordance with different individual temperaments. But it also was a step in the depersonalization of the patient into a statistic, an experimental subject rather than an individual. This happened first with humans rather than with animals: until well into the nineteenth century, the number of animals used in any one experiment was far too small to allow for statistical inference.

Vivisection in the Garden: Stephen Hales and Satire

Although Harvey had employed quantitative arguments, research on animals in the seventeenth century was largely qualitative rather than quantitative. Mechanical models for animal form and function were not necessarily mathematical models. Boyle and Hooke did not attempt to measure the amount of air they pumped out of their air pump. Fifty years later, the English clergyman Stephen Hales (1677–1761) continued their experimental program of applying the principles of physics to the exploration of life. But the laws of motion Hales employed were those Newton had elaborated, the circulatory hydraulics he explored were analogous to the movements of the planets in the heavens, and numbers and measurement were at the center of his work. Hales wrote, "Since we are assured that the all-wise Creator has observed the most exact proportions, *of number, weight, and measure,* in the make of all things; the most likely way therefore, to get any insight into the nature of those parts of the creation, which come within our observation, must in all reason be to number, weigh and measure."*

Hales had begun experimenting on animals while still a student at Cambridge University in the early 1700s. He and his friend William Stukeley (1687–1765) dissected and vivisected animals in Stukeley's college rooms. Stukeley was especially good at catching some of the many stray cats and dogs in Cambridge for use in their experimenting. Hales went on to become a Church of England clergyman, but he continued to experiment. While at Cambridge, Hales had attached a tube to a dog's artery and noted the rise and fall of blood at systole and diastole. This was the beginning of his research on blood pressure, most of which took place in the garden of Hales's home in the village of Teddington, near London, where he became vicar in 1709. He continued to use dogs but performed his research mainly on horses, because they were big enough so that Hales could easily measure their blood vessels and the rise and fall of blood pressure, as well as the force of the heart. Little was known about these topics; estimates for the force of the heartbeat were wildly exaggerated, with claims that it exerted a force of 100,000 pounds with each beat. Hales proceeded by attaching a flexible tube to an opened artery, which then connected to a long glass tube that acted as a pressure gauge. The blood rose and fell in the tube with each heartbeat and Hales measured the height, noting that the pulse rate and the output of the heart varied with stress and exertion. He also compared the blood pressure, heart rate, and cardiac output of large versus small animals, finding that blood pressure is higher in larger animals but that the output of the heart is proportionately greater in smaller ones.

Hales was not a physician, and his work did not—nor was it meant to—have any immediate medical application, yet it eventually led to the development of instruments for measuring blood pressure, as well as to the idea that blood pressure might be medically significant. Hales's interest was in hydraulics, and he employed very similar techniques in his study of plants, fixing tiny tubes to the veins of plants in order to measure the force of the flow of sap. Hales turned to plants because, he said, of the "disagreeableness" of experimenting on animals. He eventually returned to animal research.

Hales's work with horses could hardly have been a secret in a small village, and it is likely he enlisted the help of his parishioners to tie down the horses he used. There is a Monty Python grotesquerie in the image of this shy bachelor clergyman (he married at forty-three, but his wife died soon after) cutting

*Stephen Hales, *Vegetable Staticks* (1727), ed. M. A. Hoskins (London: Macdonald, 1969), xxxi.

the throats of horses in the back garden, and Hales indeed became a target for critics and satirists. While Boyle and Hooke carried out their experiments with little outside criticism, Hales was not as fortunate. Although the English public gradually accepted experimentation on humans in the form of smallpox inoculation, some members of the public were much less tolerant of experimentation on animals.

In the coffeehouses and taverns, science had become a popular topic of discussion. In the 1690s, condensed versions of the *Philosophical Transactions* of the Royal Society of London and other scientific works—in English, not Latin—appeared. All over Europe and in the American colonies, scientific literature was readily available, either in scientific journals or in the new popular magazines. Because science was only beginning to develop a specialized vocabulary, most educated people could understand the reports in the *Philosophical Transactions* or its equivalents on the continent. Science was not yet a profession, and the interested amateur like Hales was as much the norm as a paid experimenter like Hooke. The Royal Society was full of people who did not pursue science as their living. Instruments such as air pumps often formed part of the fashionable "cabinet" of curiosities kept by many aristocrats.

In this age of satire, science was often a target. In the 1670s, as we saw in Chapter 2, Thomas Shadwell's play *The Virtuoso* took aim at Royal Society experiments in the character of Sir Nicholas Gimcrack. In 1726, the Irish writer Jonathan Swift (1667–1745) published *Gulliver's Travels*. On the flying island of Laputa, Gulliver found that its intellectuals were so unworldly that they required "flappers" to slap them in the face periodically and snap them from their perpetual reverie. Laputan exiles founded the Academy of Projectors of Lagado, which bore a certain similarity to the Royal Society. Its members occupied their days with ridiculous and often repulsive experiments. While Sir Nicholas Gimcrack repeated Hooke's experiment of keeping a dog breathing by means of bellows, Gulliver witnessed an Academy member performing the operation on the opposite end of the dog. In the popular magazine the *Tatler,* a 1711 article referred to idle physicians who cut up dogs and put cats in an air pump for amusement. The same satirist referred to Colombo's experiment with a pregnant dog as if it were a current event, although it had been performed a century and a half earlier.

Hales's work on animals was widely known, especially following the publication of *Haemastaticks* in 1733. Just two years earlier, the popular *Gentlemen's Magazine* had published an article advocating vegetarianism. Now his neighbor, the poet Alexander Pope, referred to Hales as having hands "imbrued with blood." Pope had written an essay, "Against Barbarity to Animals," several years earlier, and was well known as an owner of dogs, some of whom appeared in his poems. Another poet referred to:

> Green *Teddington's* serene retreat,
> For philosophic studies meet,
> Where the good Pastor, *Stephen Hales,*
> Weigh'd moisture in a pair of scales:
> To ling'ring death put mares and dogs,
> And stripp'd the skin from living frogs
> (Nature he loved, her works intent
> To *search,* and, sometimes, to torment!)
> —Thomas Twining, "The Boat" (1740)

The standard of acceptable behavior continued to change. In the eighteenth century, the new urban culture of coffeehouses and theaters demanded new modes of behavior centered on the notion of politeness. Shadwell portrayed Sir Nicholas Gimcrack as absurd but not especially cruel, but the literary critic Samuel Johnson (1709–1784) referred in 1758 to "wretches, whose lives are only varied by varieties of cruelty; whose favourite amusement is to nail dogs to tables and open them alive."* He also asserted that physicians who performed animal experiments were thereby made cruel, the old Christian argument. Cruelty to animals increasingly was viewed as one aspect of older, unacceptable behaviors.

On his final voyage, Gulliver encountered the Houyhnhnms, horses who were culturally and morally superior to the humans, known as Yahoos, whom they governed. The Yahoos were filthy, immodest, and violent, while the Houyhnhnms were clean, trustworthy, and wise. Swift questioned what had previously simply been assumed, that humans are indeed innately superior to other animals. While his question remained unanswered, that he could ask it, even satirically, indicated that a new attitude was evolving.

Vitalism and Mechanism

Although Hales was a mechanist, by the 1730s, the mechanical philosophy was being challenged in biological explanation. Even though he used mechanical imagery, Harvey did not believe that the body operated strictly according to mechanical principle, but that there was a distinct vital principle that separated life from non-life. These "vitalist" ideas were now revived by scientists who were dissatisfied with mechanical explanations. Yet there was not a di-

---

*Samuel Johnson, *The Idler* no. 17, 5 August 1758.

The Clyster Remedy

Clysters (enemas) were a common remedy in the eighteenth century, as the illustration satirizes. In Swift's *Gulliver's Travels*, Gulliver narrates his experience on Lagado:

L'indisposition de Jocko. N.2.

I was complaining of a small Fit of the Cholick; upon which my Conductor led me into a Room, where a great Physician resided, who was famous for curing that Disease by contrary Operations from the same Instrument. He had a large Pair of Bellows, with a long slender Muzzle of Ivory. This he conveyed eight inches up the Anus, and drawing in the Wind, he affirmed he could make the Guts as lank as a dried Bladder. But when the Disease was more stubborn and violent, he let in the Muzzle while the Bellows were full of Wind, which he discharged into the Body of the Patient, then withdrew the Instrument to replenish it, clapping his Thumb strongly against the Orifice of the Fundament; and this being repeated three or four Times, the adventitious Wind would rush out, bringing the noxious along with it (like Water put into a Pump), and the Patient recovers. I saw him try both Experiments upon a Dog, but could not discern any Effect from the former. After the latter, the Animal was ready to burst, and made so violent a Discharge, as was very offensive to me and my Companions. The Dog died on the Spot, and we left the Doctor endeavouring to recover him by the same Operation.

■ Quote from Jonathan Swift, *Gulliver's Travels* (1726; reprint, Oxford: Oxford University Press, 1999), 190–91. Illustration: "L'indisposition de Jocko," ca. 1830. Courtesy of National Library of Medicine.

chotomy between mechanists and vitalists, and there were many gradations and syntheses between the two.

The basic principle of mechanism was that the organism could be explained solely according to laws of physics, that is, in terms of matter and motion. Vitalism's basic premise was that the organism was essentially different from inorganic nature. This difference was usually expressed in terms of a "vital force" not reducible to the laws of physics. In the eighteenth century and into the nineteenth, the debate between mechanism and vitalism encompassed a wide realm of investigation into the workings of the animal and human body, including generation, muscular motion, metabolism, and digestion. Both mechanists and vitalists experimented on animals to prove their assertions; indeed, because vitalists firmly believed that life and non-life were essentially different, some

The Best Experimental Subject

Researchers have experimented on themselves for centuries. Who could be a better research subject? In the early seventeenth century, the Italian physician Santorio Santorio (1561–1636) spent most of his days, for thirty years, on a large balance, or scale, allowing him to measure his intake of food and drink and his bodily discharges. He concluded that the body loses each day a quantity of fluid that he referred to as "insensible perspiration." This was an important discovery in determining how much fluid the body actually needs. A century later, another Italian, Lazzaro Spallanzani (1729–1799) investigated the digestive process, particularly the role of the gastric fluid in digestion. A French researcher had forced tame birds to swallow small, perforated tubes filled with food to investigate the digestive process, and the differing rates of digestion of different foods. Spallanzani performed similar experiments using himself as an experimental subject. He swallowed cloth bags and wooden tubes with various foods inside, which he later vomited up and studied. He also swallowed sponges to retrieve samples of stomach fluid. With

these samples he conducted *in vitro* experiments, which showed that, contrary to contemporary theories, digestion was primarily a chemical process that was accelerated by heat.

■ Santorio in his chair, woodcut, 1646. Courtesy of National Library of Medicine.

were perhaps even more likely to experiment on live animals. Cartesian mechanical philosophy had ceased to be—if it ever was—the main motivation for animal research.

The Swiss Abraham Trembley (1710–1784) made an important argument in favor of vitalism with the 1742 publication of his account of the freshwater hydra, a small and primitive animal. This so-called polyp possessed the power of regeneration. When Trembley cut it up, each piece regenerated a new polyp; life, he concluded, is inherent in living tissue and cannot be fully explained by the arrangement of its parts. Others noted that frogs and newts could also regenerate parts. Yet five years later, in 1747, the Frenchman Julien Offray de la Mettrie (1709–1751) issued his uncompromising materialist manifesto, *Machine man,* which denied any special vital or mental properties even to humans.

Trembley's account rather than La Mettrie's inspired research on the fundamental question of the nature of life. Can the basic parts of the organism be defined as living, or is only the organized being or organism so defined? The most significant contribution to this problem was made by another Swiss, the physician Albrecht von Haller (1708–1777). Haller, a professor of medicine at Göttingen in Germany, was one of the best-known physicians of his day and a strong advocate of smallpox inoculation. He performed a lengthy series of experiments on animals in the early 1750s on the quality of irritability in living tissues, defined as the ability of living tissue to respond to stimuli, usually by contraction. This quality was first noticed by Francis Glisson (1597–1677), who thought all body fibers were irritable. After Glisson's work, discussion centered on mechanical as opposed to vital explanations of irritability: Was there a way to explain irritability, and more generally muscular motion, in mechanical terms? Hales had attempted to find a hydraulic explanation, but he concluded that blood pressure had little to do with muscular motion. Haller sought instead to give a purely experimental definition, following Isaac Newton who had claimed he simply described phenomena without assigning causes.

To Haller, irritability was the ability of a muscle fiber to contract upon stimulation. It was an unconscious response of the organism, dependent not on the nervous system but on a quality inherent in muscle tissue. He classified this phenomenon on a scale of highly irritable (contracts with slight stimulation) to slightly irritable (contracts with heavy stimulation). He also identified a second property that he called sensibility, independent of irritability. Sensibility was a conscious response of the organism to stimuli, and therefore a property of tissues that have nerves. Pain was an example of sensibility. Haller reached these conclusions, first published in 1752, as a result of experiments on nearly two hundred animals, in which he stimulated various parts of the body and recorded the response. The sources of stimuli ranged from touching, to heat, cutting, and acids.

Haller argued that experiments, being repeatable, led the way to truth. He emphasized the importance of comparative anatomy in establishing the functions of the animal body: only by comparing many species can one ascertain which structures and functions are common to all and which are specific to an individual. Haller was philosophically a vitalist, and irritability was to him an example of a vital function that could not be reduced to mechanics.

From the middle of the eighteenth century onward, another school of

The Four Stages of Cruelty

In his series of engravings "The Four Stages of Cruelty," the English artist William Hogarth (1697–1764) traced the criminal career of a coachman named Tom Nero which began with torturing dogs, went on to the beating of horses, and ended with murder. The German philosopher Immanuel Kant (1724–1804) had Hogarth's engravings in front of him in 1780 when he restated the Christian position on animals. Cruelty to animals, said Kant, would damage human moral sensibility and therefore was to be avoided. But humans had no moral obligations toward animals.

■ William Hogarth, Plate IV, "The Reward of Cruelty," from a private collection.

vitalist thought emerged at the medical school of Montpellier, France. Much like the ancient empiricists, this school rejected experiment because it interrupted the flow and spontaneity of life. They argued that life was essentially different from death, and could not be determined by the fixed laws that work in physics. Haller did not attempt to find a physical, material explanation for irritability. But by concluding that experiment is a valid method because it is repeatable, he believed that nature therefore must be determined by certain fixed laws (a point of view later known as "determinism"). He did not believe life was spontaneous in the way the Montpellier vitalists argued.

"But, Can they *suffer*?"

Haller justified his use of animals in his research on irritability: "I have examined several different ways, a hundred and ninety animals, a species of cruelty for which I felt such a reluctance, as could only be overcome by the desire of contributing to the benefit of mankind, and excused by that motive which induces persons of the most humane temper, to eat every day the flesh of harmless animals without any scruple."* That Haller felt he needed any justification at all was a sign that debate about the morality of animal experimentation continued. In 1740, the *Gentleman's Magazine* published a satirical poem about air pump experiments. Five years later, a German literary journal published a defense of animal experimentation by Christlob Mylius (1722–1754), a medical student, who argued that the benefits to humankind outweighed the costs to animals.

In the year the French Revolution broke out, the English philosopher of law and political institutions Jeremy Bentham (1748–1832) enunciated a new view of the animal-human relationship. Bentham published the principles of his new "utilitarian" political philosophy in his *Introduction to the Principles of Morals and Legislation* (1789). Bentham defined utility as the property of producing good, pleasure, or happiness in an individual. The goal of government and society—both of legislation and of ethics—should be utilitarian, actively prohibiting the occurrence of pain or unhappiness. According to Bentham, humans were not the only beings susceptible of happiness and therefore within the ethical realm. Animals also qualified.

Although legal theory had relegated animals to the status of things, Bentham argued that animals could be considered interested parties in the utilitarian scheme because they are capable of happiness, and he enumerated what the interests of animals might be. He did not deny that humans were entirely

*Albrecht von Haller, *A Treatise on the Sensible and Irritable Parts of Animals* (London, 1755), 1.

The Cowpox Vaccine

Edward Jenner's cowpox vaccine generated a huge amount of publicity. James Gillray (1757–1815), a well-known caricaturist, contributed this view of vaccination in 1802.

■ James Gillray, "The Cow-Pock, or the Wonderful Effects of the New Inoculation!" London, 1802. Courtesy of National Library of Medicine.

justified in killing and eating animals: "The death they suffer in our hands commonly is, and always may be, a speedier, and by that means a less painful one, than that which would await them in the inevitable course of nature." But he denied that humans could justifiably cause animals to suffer. He compared the status of animals with the status of slaves, and Bentham was active in the burgeoning antislavery movement in England. "The day *may* come," he added, "when the rest of the animal creation may acquire those rights which never could have been withholden from them but by the hand of tyranny." The number of legs may be as irrelevant as the color of skin in assigning rights to an individual, and the Cartesian criteria of cognitive ability and speech would be equally irrelevant. The question, Bentham eloquently stated, "is not, Can they *reason*? nor, Can they *talk*? but, Can they *suffer*?"* Although Bentham's concept was overlooked in the nineteenth century, it ultimately strongly influenced

* Jeremy Bentham, *An Introduction to the Principles of Morals and Legislation* (2d ed., London, 1823), 142–43.

twentieth-century arguments about the rights of animals. He did not mention animal experimentation, however, and the argument of utility came to be used on both sides of that debate.

Jenner and Vaccination

Just at the time when Bentham wrote, inoculation for smallpox took an unexpected turn, which emphasized the close relationship between certain animal and human diseases. Inoculation had continued throughout the eighteenth century, and historians generally agree that it had an impact on reducing the number of deaths from smallpox—although there is much disagreement about the size and significance of that impact. But inoculation, as we have seen, had its drawbacks. Although the Suttons claimed that their method of inoculation produced only a single pustule on the patient, no one understood the process of infection, and inoculation was dangerous to the patient and to those around him who might contract smallpox, even from that single pustule.

Smallpox inoculation originated with folk practice, and dairymen and milkmaids had long known that if they contracted a bovine disease known as cowpox they were very unlikely to contract smallpox. Cowpox was a fairly common disease among cows, and ulcerated teats were one of its manifestations. If a milkmaid had a small cut on her hand, it would be easy for her to contract the disease when she milked an infected cow. Cowpox was usually a mild disease in humans, manifesting itself in a few pustules, often on the hands and arms, and about a week of feeling unwell. Edward Jenner (1749–1823), an English country physician, attempted to inoculate a patient with smallpox in 1778; the patient had in the past recovered from cowpox, and the inoculation was unsuccessful. During the 1780s and 1790s, Jenner collected case histories of patients who had contracted cowpox in the past and who then were found to be resistant to smallpox, either natural or inoculated. However, he had not personally seen all of these cases, and some of them had taken place many years earlier.

Based on this evidence, Jenner determined to test his theory that cowpox inoculation would confer immunity to smallpox. On 14 May 1796, Jenner performed his first "vaccination" (from *vaccinia,* cowpox, via *vacca,* Latin for cow) on James Phipps, an eight-year-old boy. Jenner used material he had taken from the arm of Sarah Nelmes, a milkmaid who was infected with what Jenner believed to be cowpox. Sarah Nelmes had never been inoculated for smallpox, but Jenner was certain that he recognized on her hand and arm the characteristic pustules of cowpox and not smallpox. James Phipps received the cowpox matter in his arm in the same way others were inoculated with smallpox mat-

ter. He felt some discomfort over the course of the next ten days but then recovered, and he became Jenner's prize example of the efficacy of vaccination.
In July 1796 Jenner variolated (that is, inoculated with smallpox, or *variola*)
young Phipps in both arms, and nothing happened. Phipps did not get smallpox. Jenner repeated this several months later, and indeed several more times
over the next few years. While Jenner's experiment appeared to have been successful, these repeated variolations may also have built up Phipps's immunity
to smallpox.

Jenner extended his experiments two years later, creating a chain of vacci-
nations from the original subject who contracted cowpox from a cow. He then
used material from that subject to vaccinate a second, and so on through several other subjects, most of them children. This "arm-to-arm" method, also
used in variolation, became the standard practice in nineteenth-century vaccination. But Jenner did not then test (or challenge, in modern experimental par-
lance) all of those vaccinated by variolation.

Nonetheless, Jenner was convinced that vaccination was an effective and
safe alternative to variolation, and he published his results in his *Inquiry into
the Causes and Effects of the Variolae vaccinae* in 1798. By referring to *vaccinia*
as *variolae vaccinae,* that is, "cow smallpox," Jenner asserted a necessary relationship between the two diseases, a relationship, his critics quickly noted, that
was purely circumstantial. Jenner further asserted that vaccination conferred
lifelong immunity to smallpox, a claim he could not prove and which was in
fact not always true. Others tested the effectiveness of cowpox vaccination,
especially Dr. William Woodville of the London Smallpox and Inoculation
Hospital, but his results only confused the issue more, since it was likely that
Woodville's vaccine—which passed through several individuals—was contaminated with the smallpox virus.

We now know that Jenner was essentially correct, although neither he nor
his contemporaries understood why. A full understanding of the action of vaccines in inducing immunity took over a century to develop, following from
Pasteur's work in the 1880s. Jenner, like the early variolators, was making a shot
in the dark. Like them, his patients were his experimental subjects, and many
of them were children. It cannot be denied that Jenner risked the lives of his
patients. It also cannot be denied that naturally caused smallpox was declared
definitively eradicated on earth in 1980. The connection between these two
occurrences cannot be easily reduced to a cost-benefit analysis.

The controversy over smallpox inoculation in the eighteenth century is relevant to many issues surrounding experimentation on humans today, particularly the notions of consent and of acceptable risk. Inoculation helped more

than it harmed, but no one had sufficient information (not even the physicians) to predict the outcome in a particular case. Jenner's vaccination minimized some risks but introduced others. While this large-scale human experiment went on, however, experiments on animals were scrutinized more and more closely, and critics expressed increasing concern about cruelty to animals and its impact on humans. All of these issues would only intensify in the nineteenth century.

4 Cruelty and Kindness

Early in 1825—almost two hundred years after Harvey had discovered the circulation of the blood—Richard Martin (1754–1834), a member of the British parliament, spoke to the House of Commons about his bill to abolish bear-baiting and other animal sports. He concluded (as reported in the official record): "There was a Frenchman by the name of Magendie, whom [Martin] considered a disgrace to society. In the course of last year, this man, at one of the anatomical theatres, exhibited a series of experiments so atrocious as almost to shock belief."* This Magendie had nailed "a lady's greyhound" to a table and then proceeded to perform the most horrific operations upon its face and body, which Martin described in graphic detail to the House, amid cries of "Shame" and "Hear, hear!" Martin's bill did not pass, however, and doubt was cast on the story of the greyhound, some saying it did not happen at all, others that the dog in question was not a greyhound but a spaniel. But in Britain, Magendie's reputation was henceforth firmly established as the French butcher, and the nascent antivivisection sentiment in Britain found a focus across the English Channel.

Who was Magendie, and why did he arouse such a furor? François Magendie (1783–1855) was part of a group of French scientists who established experimental physiology as a key science. Indeed, the very term *physiology* was coined by this group. Their goal, simply, was to find out how organisms worked, in terms of the interdependence of the various functions of a living body. Anatomy, chemistry, and physics all played roles in this investigation, but vivisection was at the center of the enterprise. In addition, they pursued their goals without reference to a more general system of physiology. On the issue of vitalism versus mechanism they remained neutral, while borrowing ideas from both sides.

Harvey, Haller, and many others had experimented on animals. What was

---

* Quoted in J. M. D. Olmstead, *François Magendie* (New York: Schuman, 1944), 140.

different about Magendie and his circle? Earlier experimenters sought to answer specific questions about animal (and human) function, questions posed by existing theories that had not been determined by experimental means. They had not established a continuous tradition of systematic experimental research, in which one experiment led to another. After Magendie, experimentation was, and continues to be, the distinguishing feature of research in physiology, no matter what the permutations of physiological theory. Where experimentation had previously been exceptional, it was now the rule.

Magendie was the beneficiary of the long line of experimenters who preceded him. But the events of the late eighteenth century in France also shaped his outlook. Over the course of the eighteenth century, medical training, led by medical schools in Leiden and Edinburgh, moved from an earlier model dominated by texts and lectures toward hands-on training that included observation of patients and dissection of cadavers. Surgery, the manual side of medicine, began to merge with the theoretical medicine of the physicians, and surgeons' skills became more valued. In France, veterinary medicine also contributed to this reformation of medical theory and practice. The founding of prestigious veterinary schools in the 1760s at Lyons and at Alfort, near Paris, established a model of professional education, and medical reformers assumed the unity of human and animal medicine. The availability of animals for research at the veterinary schools made them centers for animal experimentation, and the demands of cavalry warfare in the Napoleonic era only increased their value to the state. Many researchers spent their early careers there.

During the French Revolution, which began in 1789, medical and surgical training focused on the hospitals. Revolutionary ideology discounted the misguided past and emphasized empirical knowledge and the value of analysis rather than theorizing in medicine and the sciences. The resulting "French clinical school" of the first third of the nineteenth century made Paris the acknowledged center of medical authority. While physiologists sometimes complained that medical reliance on clinical observation and autopsy ran counter to their own interest in experimentation, in fact medicine provided physiologists with necessary skills and with interesting problems to study.

Magendie trained as a surgeon, and began his research career at Alfort. He and his generation were inspired by the work of Xavier Bichat (1771–1802), a surgeon and anatomist who had died at the early age of thirty. In his *Discourse on the Study of Physiology* (1798), Bichat set forth the criteria for a new science. While mindful of the uncertainties involved in experiments on live animals, Bichat combined autopsy, anatomy, and vivisection into a powerful research

program. Because vital forces were inherently unstable, animal experimentation required strict rules and careful procedures. He established general rules
for experimentation that became the basis of the new science of physiology:

- compare the experimental animal with a normal one as a control
- eliminate "accidental interferences": that is, control the experimental
  environment as much as possible
- repeat the procedure to ensure that the same result will be obtained
- carefully examine the state of the animal before and during the
  experiment

Bichat is best known for his tissue doctrine, which stated that the organism or even the organ was not the fundamental unit of analysis in the body and
that yet simpler organic elements existed, which he identified as twenty-one
tissues. In the same way that chemistry in the 1790s developed into a science
of elementary bodies with the concept of the chemical element, so anatomy,
said Bichat, would be a science of elementary tissues, the study of these simple
organic "elements" and the structures they form. The principle of life was at
the level of the tissues. Bichat's tissue doctrine had two effects on research in
the nineteenth century. First, it stimulated the continued search for even more
fundamental levels of analysis than tissues, resulting in the development of cell
theory in the 1830s and 1840s. Second, it established the experimental method
as the most fruitful method for physiological research by establishing the idea
of organic function as uniquely biological. Bichat's definition of life as "the
totality of functions which resists death" bypassed the vitalism/mechanism
question and focused research on the organism.

Bichat believed that all of the tissues of the body possessed the principle of
life, and to him this implied that all tissues possessed some degree of sensitivity, feeling, and movement. Magendie and his circle, while retaining Bichat's
emphasis on experiment, disagreed with this view. In their experimental work,
they rejected Bichat's emphasis on the body as an organic whole and concentrated on isolated systems. Julien Legallois (1772–1814) argued that life was centered not in the body as a whole but in the spinal cord. To support this view,
he separated a nerve from its connection to the spinal cord and showed that
the affected part was thereby deprived of sensation and motion. He also showed
that even parts thought to be merely "irritable" by Haller had in fact some connection to the spinal cord, although he could not entirely explain why the heart
could continue to beat in an animal whose spinal cord was destroyed.

Legallois wrote, "Experiments on living animals are among the greatest
lights of physiology. There is an infinity between the dead animal and the most

feebly living animal."* This statement might serve as Magendie's motto. Of all of his contemporaries, Magendie remained the most devoted to experimentation. He viewed himself as the Newton of physiology, the founder of a new science, and he detailed its principles in his new journal of experimental physiology. Magendie's goal was to make physiology a science on the level of certainty of Newtonian physics. He felt that physiology had been damaged by the speculative tendencies of vitalism, and like Newton, he claimed he would merely observe and not draw conclusions about causes. Any function, he said, was the sum of actions of a number of organs and did not depend on any particular vital property. The fundamental distinction was simply between living and non-living, and what we call vital phenomena may have physical or chemical causes of which we are not yet aware. Experimentation, by establishing the anatomical organization of functions and their mode of action, could reveal the laws by means of which vital phenomena operate. Dissection of the dead could not do this.

Like Legallois, Magendie focused his research on the nervous system. In his first experiments, he tested the action of a poison on several dogs and other animals. The poison was introduced on a sliver of wood into the thigh of the animal, who shortly thereafter began to experience symptoms of poisoning, including paralysis. Magendie concluded that the poison was carried by the circulation to the spinal cord, a conclusion he tested by introducing the poison into other parts of the body. Injection into parts well supplied with blood vessels led to rapid absorption of the poison; in other parts of the body it acted more slowly, and ingested with food it acted even more slowly. Direct injection into the arteries produced rapid action. Experiments to separate the spinal cord from the brain and to destroy part or all of the spinal cord showed that the cord, as he claimed, was the seat of action for the poison.

This series of experiments demonstrates Magendie's techniques and philosophy. He attacked a single question from a number of angles, testing and retesting systematically, never satisfied with the obvious conclusion. Even facts that were apparently well known required new tests, new experiments. He did not view the organism as a whole but as a series of individual systems. He localized feeling or sensibility in the nervous system and not in individual organs. In another experiment on the action of an emetic drug, he showed that it acted via the nervous system, not by the direct action of the stomach alone. By repeating experiments and testing alternate explanations he used many animals,

* J. J. C. Legallois, *Expériences sur la principe de la vie* (Paris, 1812), xxiii, quoted in J. E. Lesch, *Science and Medicine in France: The Emergence of Experimental Physiology, 1790–1855* (Cambridge: Harvard University Press, 1984), 89.

and many different kinds of animals. Many of these experiments were undoubtedly painful to the animals.

Although Magendie occasionally used opium to anesthetize his experimental animals, his experiments on the spinal cord and the roots of the nervous system required that the animal be fully alert. In the early 1820s, a detailed and painstaking series of experiments on puppies and later on other animals demonstrated that each nerve had two roots (known as posterior and anterior, from their position) on the spinal cord, and that the posterior roots controlled sensibility, while the anterior roots controlled motor skills. In other words, by cutting one nerve the dog could walk but not feel; by cutting another the dog could feel but not walk. The organism as a whole could not feel; only the nervous system could. Life resided in the brain and spinal cord, and feelings and emotions were mediated by the brain. Not since Galen had such systematic analysis been undertaken of the nervous system, with the extensive vivisection this entailed. Like Galen, Magendie often appeared to his contemporaries and to modern readers to be cruel and unfeeling toward his experimental animals.

The Beginnings of Antivivisection in Britain

Richard Martin was one of many Britons repelled by Magendie's experiments. A British visitor to France in 1817 commented with disapproval of the "mania for vivisections" of Magendie and his circle, but the Frenchman's experiments on the nervous system aroused much scientific interest in Britain. Magendie was probably inspired in his own research by a demonstration given at Alfort in the early 1820s by an assistant to the English researcher Charles Bell (1774–1842). Indeed, the differentiation of nerve functions is known as the "Bell-Magendie Law." When Magendie came to London in 1824, therefore, he was well known in scientific circles, and the differences between his style of research and Bell's highlighted the broader differences between British and French attitudes toward animals. Where Bell emphasized the anatomy of cadavers and dead animals, supplemented by vivisection (often on previously "stunned" animals, which would limit their ability to display sensibility), Magendie spoke disparagingly of the inadequacy of anatomy alone and vivisected without hesitation. "Soyez tranquille" (be still), he said to a restless dog during his London demonstrations, adding jokingly, "Il serait plus tranquille s'il entendait français" (he would be calmer if he understood French). This comment was repeated in the British press as an example of his heartlessness.

How had these differences in attitude arisen? Why was Magendie lauded in France while he and the British researcher Marshall Hall (1790–1857) were vilified in Britain? One answer is that vivisection provided a touchstone for a multitude of ideas and emotions in an age of rapid and turbulent change in politics, culture, and society, especially in Britain. The 1820s witnessed the disillusionment of the Romantics whose hopes had been kindled by the revolutionary ideals of 1789. While Mary Shelley's parents, the political radicals Mary Wollstonecraft and William Godwin, believed that science and reason could remake the world, in *Frankenstein* (1818), their daughter vividly refuted their claim: science and reason, pressed to its limits, led to horror, despair, and death. Victor Frankenstein built his creature of body parts from stolen corpses, but did not then assume the responsibility for his creation. His neglect of and cruelty toward the creature leads eventually to his own death. Mary Shelley called Frankenstein "the modern Prometheus," referring to the figure in Greek mythology who stole fire from the gods and was punished for his hubris, for his overweening pride.

France remained overwhelmingly rural in the 1820s, but industrialization had already led to widespread urbanization in Britain, so that cities such as Manchester and Leeds, which had been sleepy backwaters a century earlier, were now booming metropolises. Romantic poetry in Britain idealized the countryside, not only the wilderness, but also the pastoral landscape of farms and fields that was rapidly disappearing. Alongside the old social order with its clear boundaries of rank and station arose a new society of industrial entrepreneurs and restless workers who demanded a share of power from the traditional aristocracy. The Great Reform Bill of 1832, while hardly radical, gave parliamentary representation to the new urban metropolises and extended voting rights for the first time in four centuries. Meanwhile, evangelical clergy pondered the morality of slavery, which was finally abolished in 1838. Traditional structures of society seemed to be collapsing, along with traditional values. New urbanites, as well as old gentry, longed for a return to an ideal rural past of deference and paternal kindness.

What did these changes mean for animals? Animals, especially domestic animals and above all dogs, provided the new urbanite with a connection to the rural past. Petkeeping, long an aristocratic privilege, became established among the middle classes at the end of the eighteenth century. The proliferation of dog (and cat) breeds in the nineteenth century helped distinguish middle-class pets from working animals and from the mutts of the lower classes. In addition, the fabled loyalty and faithfulness of dogs stood in contrast to the instability of modern life.

In 1800, a bill in the House of Commons opposing the "savage custom of bull-baiting" was greeted with derision and defeat, and Mary Wollstonecraft's 1792 *Vindication of the Rights of Women* was parodied by a *Vindication of the*

The Anatomy Act

Dr. Frankenstein stole bodies to make his monster. He "dabbled among the unhallowed damps of the grave, [and] tortured the living animal." In the illustration, the famous London anatomist William Hunter is caught in the act of stealing a corpse for dissection. But a year after the second edition of *Frankenstein* appeared in 1831, grave robbing was no more in Britain. The Anatomy Act of 1832 gave to the anatomists the bodies for which they clamored. Previously, anatomists and surgeons had legal access only to the bodies of executed criminals. The "Murder Act" of 1752 had made dissection part of the punishment for murder. Surgeons sometimes struggled with the families of the executed for possession of the precious corpse. But the growth in importance of pathological anatomy, and of anatomy as part of medical training, meant that this supply of bodies was inadequate for the purposes of the anatomists. Grave robbing had resurged in

the eighteenth century. In Paris, Bichat was known for his skills in cemetery raiding. The case of Burke and Hare in Edinburgh in 1828 brought the issue to the fore, and arguments in Britain for and against the proposed parliamentary act played on the complicated relationship between human and animal bodies, life and death. The Anatomy Act stated that the bodies of the destitute who died in government-run workhouses and hospitals would be available for anatomists. From being a punishment for murder, dissection became a punishment for being poor. Thereafter, the fear of dying destitute and in hospital haunted generations of the poor.

■ Quote from Mary Shelley, *Frankenstein* (1818), reprint in *The Essential Frankenstein,* ed. Leonard Wolf (New York: Penguin, 1993), 81. Illustration: William Austin, "The Anatomist Overtaken by the Watch . . . Carrying off Miss W - - - ts in a Hamper," 1773. Courtesy of National Library of Medicine.

Rights of Brutes: what could be more absurd? But in 1822, Richard Martin succeeded in guiding his bill forbidding cruelty to farm animals through the House of Commons (the act was later amended to include dogs and cats). In this he was aided by the antislavery evangelical clergyman William Wilberforce as well as by a mass of petitions, mostly from urban areas.

The defeat of Martin's bill in 1825, in the course of which Martin had decried Magendie's barbarous science, was only a momentary setback. In 1835 Martin's Act of 1822 was amended to cover animal sports. In 1824 the first animal protection society was founded in London, the Society for the Prevention of Cruelty to Animals, or SPCA. Queen Victoria's sponsorship a decade later allowed "Royal" to be prefixed to the SPCA. The RSPCA, whose membership was largely urban and upper class, focused on those animal sports Martin had sought to outlaw in 1825, such as bull- and bear-baiting. These were sports of the lower classes—the RSPCA did not address the upper-class practice of fox hunting—and it was argued that banning these sports would help to civilize the naturally violent lower orders. To be humane was a sign of civilization, and following the old argument, which made Wilberforce and Martin natural allies, humans who were cruel to animals would also be cruel to their fellow humans.

Martin was a founding member of the SPCA, and as his criticism of Magendie shows, these early animal protectionists opposed vivisection, although they had few examples of it to point to in Britain. A steady flow of letters in the popular press as well as in medical journals nonetheless kept the topic current. In 1829, an editorial in the *London Medical Gazette* recalled the popular outcry over Magendie's visit five years earlier in arguing in support of a bill that would give anatomists access to the cadavers of the poor for dissection. A year earlier, the trial of Burke and Hare in Edinburgh, who murdered indigents and then sold their bodies to Dr. Robert Knox of the Medical School for dissection, highlighted the need of scientists for cadavers but also revealed the anatomists as heartless and unscrupulous, recalling once more their association with the hangman.

Foremost among British vivisectors was Marshall Hall, an Edinburgh-trained physician who kept a laboratory in his house while also continuing a medical practice. In the 1820s he began to investigate the effects of blood loss. His first major paper was based on observations of his patients, but by the late 1820s he was experimenting on dogs. Since bleeding continued to be a popular therapy, this research had considerable practical significance.

Perhaps inspired by Magendie, by 1830 Hall was studying the interaction between the nervous system and the circulation of the blood. In 1831, the year of fierce parliamentary debates over the Anatomy Bill as well as the year in which the second edition of *Frankenstein* reminded readers of the necessary limits of science, Hall published *A Critical and Experimental Essay on the Circulation*. In this work he described experiments on fish and frogs that included crushing the brain and spinal cord as well as other parts of the body to observe the effects on the circulation. Mindful of antivivisection criticisms, Hall called for the formation of a "society for physiological research," which would regulate animal experimentation, for he said, "every experiment . . . is necessarily attended by pain or suffering of a bodily or mental kind."* He proposed five guidelines for the performance of animal experimentation:

1. experiments should be absolutely necessary, and all alternatives should be explored
2. experiments should have clear and attainable objectives
3. experimenters should avoid unwarranted repetition, which includes being aware of the work of past experimenters
4. the least possible pain should be inflicted (in this age before anesthesia, Hall recommended using "lower" animals such as frogs and fish or newly dead animals)
5. experiments should have witnesses to certify results and lessen the need for repetition

The society never materialized, and despite these moderate recommendations, Hall, who continued to experiment, became a focus for antivivisection feeling in Britain in the 1830s and 1840s. Anti-Magendie and more general anti-French sentiment also abounded. In 1847 Hall replied to an extended critique of his work, leading to a series of letters and replies, including an antivivisection pamphlet published by the RSPCA. Hall continued to work on his own, with no official position or support, and when he died in 1857, he was still one of very few experimental physiologists in Britain.

The Introduction of Anesthesia

In 1846, in the midst of the debates between Hall and his opponents, the first operation that successfully employed ether anesthesia was performed in Boston, Massachusetts. The introduction of anesthesia ultimately had a profound impact on perceptions of pain, changing the relationship between doctor and patient, as well as between experimenter and animal. But its early use also

---

*Quoted in Diana Manuel, "Marshall Hall (1790–1857): Vivisection and the Development of Experimental Physiology," in N. A. Rupke, ed., *Vivisection in Historical Perspective* (London: Croom Helm, 1987), 85.

revealed strongly held attitudes about the capacity for pain of different levels of society, including animals.

Pain has meant different things in different cultures, and as we have seen in Descartes's work, pain and suffering were not necessarily considered to be the same thing. Pain was a constant companion in premodern society. Today, for everyday pain such as headaches, we can swallow a pill and know that our pain will be relieved soon. Before the twentieth century there was no such certainty, and a stoic demeanor seemed the best way of dealing with chronic minor aches. Christian theology, which viewed earthly suffering as a necessary prelude to heavenly bliss, contributed to this outlook. Because pain was viewed as a symptom of illness, the relief of pain was only incidental to curing the illness. Indeed many premodern medical practices, such as bleeding and cauterization, caused more pain. Pain, like fever, was also seen as nature's reaction to a bodily crisis, and therefore suppressing it could be harmful to the patient. The pain of surgery, therefore, was necessary to the healing process, and some claimed that more painful operations had greater success than less painful ones. In a twist on Descartes, some physicians argued that the ability to feel pain was what made us truly human.

Opium, in the form of liquid laudanum, was available, as was alcohol in various forms, but neither of these was entirely satisfactory as an anesthetic agent: dosages were unpredictable, and the side effects undesirable. The quantity of either drug needed to produce total stupefaction was so great as to be dangerous, and experimental techniques to evaluate the relationship between a lethal dose and an effective dose had not yet been developed. The English writer Frances Burney took a glass of sherry before she endured a mastectomy in 1811; as demonstrated by her horrifying account of the operation in a letter to her sister Esther, the alcohol had little effect on either her pain or her terror.*

Morphine, the active ingredient in opium, was isolated at the beginning of the nineteenth century. A number of experiments on animals, especially dogs, followed to determine its mode of action: was it a stimulant, a depressant, a poison? By the late 1820s, morphine was used in hospitals not only for pain relief but for a variety of other ailments, and its addictive qualities were beginning to be noted. The existence of new substances to relieve pain, together with the new status afforded the sensitive character, meant that pain relief came to be a much more important part of medical care over the course of the nineteenth century.

* Joyce Hemlow, ed., *The Journals and Letters of Fanny Burney,* vol. 6 (London: Oxford University Press, 1975), 596–615.

In contrast to opium or alcohol, ether and its near companions nitrous oxide and chloroform acted quickly. Nitrous oxide in low concentrations gave only an incomplete state of anesthesia, while chloroform was highly potent. When inhaled in the form of a vapor, ether put the subject into a state of unconsciousness resembling sleep from which the subject was not aroused by painful stimuli. The medical profession was divided on the topic of anesthesia, and some physicians worried that the patient experienced pain but forgot it. Inhaled forms of anesthesia quickly grew in popularity. Dr. James Young Simpson in Edinburgh employed chloroform for a difficult childbirth in 1847. Soon its use for this purpose spread, despite critics who warned that pain in childbirth was necessary to the process of giving birth, both because it stimulated contractions and because it provoked maternal love for the new child. (Chloroform actually relaxes the uterine muscles, thus diminishing the effectiveness of contractions.) Others went so far as to declare that pain in childbirth was a necessary part of the punishment for Eve's transgression.

Animal experiments, especially in France, were conducted to find how ether and other anesthetic gases worked: Was loss of consciousness caused by the chemical or by lack of oxygen? How long could ether anesthesia be prolonged without ill effects to the patient? Cases of death by anesthetic asphyxia were not uncommon. Magendie objected to the use of ether on two grounds: because its mode of action was unknown, using it on patients amounted to human experimentation, which he could not countenance; and he found it unethical to operate on an unconscious person, who could not assess the surgeon's performance during the operation and who was therefore at his mercy.

Like smallpox inoculation, the benefits of anesthesia seemed to outweigh its risks. However, medical men believed that the capacity for pain differed widely among humans, on the basis of "age, sex, and temperament," according to Magendie,* but also race and social class. This categorization by the ability to feel pain provided a scientific affirmation of the ideal social order and reflected the actual power relationships in Western society. Women, of weaker nerves, were generally believed to be more sensitive to pain than men, and the upper classes felt pain more acutely than the lower classes. Apart from temperamental differences, which tended to differentiate classes, the hard labor of the poor inured them to pain. Different races—a rubric that included in nineteenth-century America the Irish and other immigrant groups as well as Africans and Native Americans—also felt pain differently, whites being deemed the most sensitive group. The American gynecological surgeon J. Marion Sims

* François Magendie, *Elementary Compendium of Physiology* (trans. 1824), 98, quoted in Martin Pernick, *A Calculus of Suffering* (New York: Columbia University Press, 1985), 148.

performed experimental surgery only on black slave women because, he said, white women felt pain more.

Children occupied an ambiguous status in this hierarchy of pain. Some physicians believed that children, like women, were more sensitive to pain. But others argued that young children, lacking reason, resembled animals in their relative insensitivity, an argument very close to Descartes's, although he denied this implication for children. Even today there is considerable argument about the capacity of newborns to feel pain, and the use of anesthetic agents in such operations as circumcision was, until quite recently, much debated.

Animals were at the bottom of this hierarchy, below even the least sensitive human. Many believed that animals simply did not feel pain with the same intensity as humans did, although, mirroring the human hierarchy, domestic animals were believed to be more sensitive than wild ones. Since operations on certain classes of humans continued to be performed without anesthesia until late in the nineteenth century, it is not surprising that the use of anesthesia on experimental animals was inconsistent. It was employed—as it sometimes was in surgery in humans—as much to keep the animal still as to relieve pain. Paralytic agents such as curare kept the animal still but did not relieve pain. While the introduction of anesthesia would, it seems, have eliminated a major objection to vivisection—that it was painful—in fact, it raised more questions than it resolved. It freed the consciences of some scientists to pursue animal experiments to which they had previously objected on the grounds of cruelty, and therefore led to an increase in invasive experiments. One antivivisectionist commented, "I am inclined to look upon anesthetics as the greatest curse to vivisectible animals."* It also forced antivivisectionists to examine the bases of their beliefs.

Claude Bernard and the Defense of Experimentation

When Magendie died in Paris in 1855, the French recognized him as one of the great scientists of the century. Unlike Hall, who had never won a university position, Magendie held a chair at the elite Collège de France, which he bequeathed to his most promising pupil, Claude Bernard (1813–1878). Bernard continued Magendie's research program and assumed leadership in the campaign to establish experimental physiology as a distinct science. At Bernard's death, the methods of experimental physiology—centered on animal experimentation in the laboratory—had become the favored means of gaining knowl-

*George Hoggan, *Anaesthetics and Lower Animals* (London, 1875), quoted in Richard French, *Antivivisection and Medical Science in Victorian Society* (Princeton: Princeton University Press, 1975), 68.

Anesthesia

Anesthesia proved to be particularly fertile territory for self-experimentation. The chemistry of gases began to be known at the end of the eighteenth century, with the isolation of oxygen and the discovery of several new gaseous compounds. Among these was nitrous oxide. The chemist Humphry Davy (1778–1829) tested nitrous oxide on himself and his friends (who included the romantic poets Samuel Taylor Coleridge and Robert Southey) in the early years of the nineteenth century. He noted both its anesthetic effects and the pleasurable rush he called "the thrilling." But nitrous oxide did not come into use as an anesthetic until many years later, and then as a supplement to more potent agents. Many of the early developers of surgical anesthesia, including the dentists

William Morton and Horace Wells and the physicians James Young Simpson and Crawford Long, first tried the drugs on themselves. Morton tested ether on himself in his office. Wells experimented with nitrous oxide and later became addicted to chloroform, ending his life as a suicide. Long first encountered ether as a recreational drug in the early 1840s and gradually realized its anesthetic quality. Simpson, the Edinburgh obstetrician, tried a number of inhalants upon himself (including acetone and benzene) before settling on chloroform as the best anesthetic.

■ "Sir J. Y. Simpson and two friends, having tested chloroform on themselves, lying insensible on the floor around a table." Pen and ink drawing, ca. 1928. Courtesy of Wellcome Trust Medical Photographic Library.

edge about the body. While Bernard made many significant discoveries about the body, this section will focus on his public role as spokesman for experimental physiology, a role that he made explicit in his *Introduction to the Study of Experimental Medicine* (1865) and that made him a target for antivivisectionists. With Bernard, the laboratory rather than the clinic or the dissecting table became the primary site for learning about the body and its functions. It also became the place students came to learn about the body, a site for training as well as for research.

Bernard intended to be a writer, but after the failure of some early works he took the advice of one of his critics and found another profession. Entering medical school in 1834, he began to work with Magendie five years later. Focusing first on digestion, he combined chemical analysis with animal experimentation and showed the interconnection of the nervous system and individual organs in the process of digestion and metabolism. In his work he defined the scope of experimental physiology: he was not concerned with the nature of life but with the experimental resolution of vital phenomena. The nature of life was unknowable, but the specific circumstances of a particular phenomenon could be experimentally determined. Although living beings are complex and changeable as a whole, their individual operations, such as respiration and digestion, followed laws and could be analyzed experimentally. Bernard presented his experimental results as an example of his philosophy of physiological determinism, which stated that experiments always produced replicable results and yielded identical outcomes.

Bernard believed that only experimental physiology could provide new information about the body, which could then be applied to human medicine. Although he had a medical degree, Bernard never practiced medicine and was scornful of medicine as a science. Its reliance on observation of sick patients and the anatomy of dead bodies could not give information about vital processes. This required the active intervention of vivisection. Nor could chemistry alone unlock the secrets of the body, although Bernard employed chemical analysis in his research. Bernard's new science combined chemistry and analysis of tissues and cells with vivisection. Animals were to serve as proxies for humans, and the goal, which ultimately eluded Bernard, was to construct a general physiology for the entire animal kingdom. The aim of experimental physiology, he said, was "to conquer living nature, act upon vital phenomena and regulate and modify them."*

*Claude Bernard, *Principes de médecine expérimentale* (Paris, 1947), 285, translated and quoted in William Coleman, "The Cognitive Basis of the Discipline: Claude Bernard on Physiology," *Isis* 76 (1985): 56.

Alexis St. Martin's Stomach

In 1822, a young French-Canadian voyageur in northern Michigan named Alexis St. Martin received a gunshot wound in his side. St. Martin's wound healed but did not close, leaving an opening into his stomach. This gastric fistula provided a perfect window into the mysteries of the digestive system, and his doctor, William Beaumont, decided to perform experiments with him. St. Martin moved into Beaumont's home as his servant, and Beaumont began regularly to poke bits of food into St. Martin's stomach. He also took samples of gastric fluid, and frequently inserted a long thermometer deep into St. Martin's stomach. All of these activities caused some distress, but the thermometer was especially painful. Between 1825 and 1833 Beaumont performed hundreds of experiments on St. Martin, which he published in articles and in a book, *Experiments and Observations on the Gastric Juice, and the Physiology of Digestion* (1833).

St. Martin was not always a willing participant in these experiments, periodically running away from Beaumont and returning home to Canada. St. Martin signed a contract with Beaumont in 1832 that bound him to the doctor for a year, in exchange for room and board and a sum of money, and he returned home permanently at the end of the year.

Beaumont's experiments on St. Martin made both of them famous. Physicians and scientists praised Beaumont, but few expressed any interest in St. Martin's welfare. After 1833, Beaumont frequently offered the impoverished St. Martin money to come back and be experimented

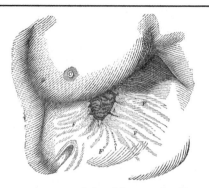

on, and was puzzled and frustrated at St. Martin's refusal. Beaumont did not coerce St. Martin, and indeed treated him well. Contemporaries found St. Martin's case to be far more defensible ethically than experimentation on animals, and no one mentioned St. Martin's rights in the matter. Some twentieth-century authors have even credited Beaumont with developing ethical principles of human experimentation, which include voluntary consent and lack of distress to the experimental subject. But St. Martin's relationship with Beaumont was more ambiguous than these ideals imply. St. Martin's financial need and lower social status put him in a dependent relationship with Beaumont, which was only partially voluntary, and Beaumont clearly put the interests of science before the interests of his experimental subject. To Beaumont and his peers, the master-servant relationship between the doctor and the fur-trapper was a natural consequence of their differing social stations.

■ St. Martin's fistula, from William Beaumont, *Experiments and Observations on the Gastric Juice, and the Physiology of Digestion* (1833), 25. Courtesy of Biomedical Library, University of California, Los Angeles.

Claude Bernard admired William Beaumont's work with Alexis St. Martin, and in his *Introduction* he justified human experimentation in certain circumstances:

> It is our duty and our right to perform an experiment on man whenever it can save his life, cure him or gain him some personal benefit. The principle of medical and surgical morality, therefore, consists in never performing on man an experiment which might be harmful to him to any extent, even though the result might be highly advantageous to science. . . . For we must not deceive ourselves, morals do not forbid making experiments on one's neighbor or on one's self; in everyday life men do nothing but experiment on one another. Christian morals forbid only one thing, doing ill to one's neighbor. So, among the experiments that may be tried on man, those that can only harm are forbidden, those that are innocent are permissible, and those that may do good are obligatory.*

He left the judgment of what was harmful or "innocent" to the experimenter.

Bernard's advocacy of experimental physiology was based on intellectual conviction and on personal circumstance. Despite being Magendie's right-hand man, he resembled Marshall Hall in having no official position or support for part of his career. Even after he was appointed to Magendie's chair, his laboratory space and resources were far less than he wished, especially in comparison to German scientists, whose large, well-equipped labs could support many students and researchers. Bernard's marriage to Marie-Françoise Martin in 1845 was one of convenience, for his wife was wealthy and helped to support his research. But the marriage was unhappy and Bernard and his wife soon became estranged. Later in life, she and their daughters became ardent antivivisectionists, much to Bernard's dismay.

Bernard's *Introduction to the Study of Experimental Medicine* was a manifesto for a new science as well as a personal plea for additional support from the French government. It also served as a wake-up call for French science. In the 1830s the Paris clinical school had been the model of modern science, but by the 1860s the German universities had usurped that position. The highly competitive German states poured money into their universities, creating laboratories and research institutes that were the envy of the world. One of the most important aspects of the German model was the role of the laboratory in training the next generation of scientists. When Germany unified into a single state in 1870—and shortly thereafter handed France a humiliating defeat in the Franco-Prussian War—it seemed an unbeatable power, and British and Amer-

---

*Claude Bernard, *An Introduction to the Study of Experimental Medicine* (1865; trans. 1927; reprint, New York: Dover, 1957), 101–2.

ican scientists as well as French scientists urged their governments to help them catch up. In the United States, the German model inspired the organization of the medical school of the Johns Hopkins University in Baltimore, founded in 1891, which became in turn the model for the reorganization of American medical education after 1900.

Although Bernard admired German success, his approach in the *Introduction to the Study of Experimental Medicine* differed significantly from the more mechanistic Germans, who assumed that physiology could be reduced to chemistry and physics. Bernard insisted on the uniqueness of life and the uniqueness of physiology among the sciences. Vital phenomena followed laws—otherwise experiment would be useless—but could not be entirely explained in terms of chemistry and physics. His model of experimentation began with observation and moved on to the creation of a hypothesis and experiments to test the hypothesis. He constantly returned to experimentation as he revised his hypotheses. Bernard viewed this as a highly rational process, but intuition also played a role in the development of a hypothesis.

Much of the book is devoted to vivisection: methods, choice of animals, and justification. "We shall succeed in learning the laws and properties of living matter only by displacing living organs in order to get into their inner environment," he wrote. "To learn how man and animals live, we cannot avoid seeing great numbers of them die, because the mechanisms of life can be unveiled and proved only by knowledge of the mechanisms of death."* Unlike Hall, who argued that the use of "lower" animals was to be preferred, Bernard frequently experimented on dogs and insisted that the more closely the animal resembled humans, the more useful it was to the experimenter (he refused to experiment on primates, however; see Chapter 6).

While Bernard was well aware of anesthesia and its uses—he wrote a treatise on the actions of anesthetics, based on extensive investigation on animals—he barely mentioned its use in the *Introduction*. He used anesthesia at times, but more to calm an animal than to relieve its pain. Manifestation of pain was part of the process of experiment, and did not call for special recognition or relief. Bernard saw himself as the successor to Galen and Harvey (his textbook on physiology has been called an updated version of Galen's *On Anatomical Procedures*), and he shared Galen's contempt for rivals and critics. The scientist was a special being, immune to the "cries of people of fashion" as well as to those of the animal itself; he saw only his scientific goal, and should be judged only by his fellow scientists. We have the right to vivisect animals, Bernard

---

* Bernard, *Introduction*, 99.

wrote, "wholly and absolutely."* It is not surprising that he succeeded Magendie as a target of antivivisectionists.

Frances Power Cobbe Fights the Antivivisection Cause

By the middle of the nineteenth century, a gap had opened between much of science and the nonscientific public. Bernard firmly closed the doors of his laboratory, and the public demonstrations of Magendie were a thing of the past. Also serving to exclude the public was the rising specialization of science, with a proliferation of journals written in increasingly technical language. Yet in his *Introduction,* Bernard deliberately addressed the public and challenged the antivivisectionists. While the French Société protectrice des animaux (Protective society for animals), founded in 1846, had formed a committee to investigate vivisection in 1860, British protest against vivisection had been far stronger.

Lacking local targets, animal protectionists in Britain continued to look overseas. In 1846, a clergyman made the first of what would be many protests to the French government over the treatment of horses at the veterinary school at Alfort. This issue came to the attention of the RSPCA in 1857, the same year in which Queen Victoria gave birth to her youngest child under chloroform anesthesia. Horses at Alfort were used for surgical training—not research— and were subjected to repeated operations without anesthesia. A series of letters in both the popular and medical press in Britain in the late 1850s and early 1860s uniformly condemned the practice, and the RSPCA organized a delegation to the Emperor Napoleon III in 1861 that demanded an end to vivisection at the veterinary schools. A French commission, of which Bernard was a member, recommended in 1863 that vivisection continue, but under more controlled conditions, including the use of anesthesia. The French Academy of Medicine, deeply resenting English interference, indignantly rejected the commission's findings.

The RSPCA's campaign against the French brought the cause of antivivisection to the attention of Frances Power Cobbe, who would become the driving force of the British antivivisection movement. Cobbe (1822–1904) embodied many of the characteristics and contradictions of the nineteenth-century movement. An upper-class Englishwoman, she worked as a journalist, writing on religious, feminist, and philanthropic topics. Forceful and energetic, she cultivated a wide range of friends and supporters from the clergy and aristocracy as well as from the literary world, and she produced a flood of influential letters and articles for the mainstream press.

* Bernard, *Introduction,* 102.

The True Vivisector

Thousands of copies of Frances Power Cobbe's antivivisection pamphlet *Light in Dark Places* (1888) were distributed for free. The illustration, from Claude Bernard's textbook of physiology (*Leçons de physiologie opératoire,* 1879), appeared in Cobbe's pamphlet, accompanied by inflammatory text by the Russian physiologist Elie de Cyon (1843–1912):

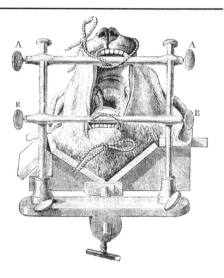

The true vivisector must approach a difficult vivisection with the same joyful excitement, and the same delight, wherewith a surgeon undertakes a difficult operation, from which he expects extraordinary consequences. He who shrinks from cutting into a living animal, he who approaches a vivisection as a disagreeable necessity, may very likely be able to repeat one or two vivisections, but will never become an artist in vivisection. . . . The pleasure of triumphing over difficulties hitherto held insuperable is always one of the highest delights of the vivisector. And the sensation of the physiologist, when from a gruesome wound, full of blood and mangled tissue, he draws forth some delicate nerve-branch, and calls back to life function which was already extinguished—this sensation has much in common with that which inspires a sculptor, when he shapes forth fair living forms from a shapeless mass of marble.

■ Frances Power Cobbe, *Light in Dark Places* (1888; reprint, London: Victoria Street Society, 1889), 15–17.

Cobbe was in Florence in 1863, and led a protest by the English community there against the work of Moritz Schiff (1823–1896), a professor at the natural history museum. Acting on hearsay evidence, Cobbe concluded that Schiff was not only experimenting on animals but doing so with unusual cruelty. She forthwith wrote up a petition, obtained almost eight hundred signatures from Britons, other foreigners, and Florentine aristocrats, and sent off a letter of protest to the London *Daily News.* She sent the petition to Schiff. He responded angrily in local newspapers, denying that any animals suffered unduly, assuring critics that they used anesthesia, and arguing for the scientific necessity of vivisection. British newspapers took far more interest in the issue than did Italian ones, and the case, like earlier protests against French practices, displayed the conviction that cruelty to animals, especially dogs, cats, and horses, was simply un-English.

In this era of anesthesia and heightened consciousness of pain, experiments that deliberately caused pain were an obvious focus for antivivisectionists. But pain was not the only issue. Evolutionary theory, recently enunciated by Charles Darwin in *On the Origin of Species* (1859), indicated that humans and animals shared common ancestors. Perhaps more important for urban Victorians, animals represented a connection with nature and a rural past that was rapidly disappearing, while science foreshadowed a modern world of ceaseless change and instability. Cobbe's allies, the clergy and aristocracy, had a particular interest in preserving older values. Yet as an independent, self-supporting, and unmarried woman, Cobbe also represented new and competing values.

Cruelty may have been un-English, but by 1870, British physiologists began to adopt continental models of experimentation, including vivisection. The publication of John Burdon Sanderson's *Handbook for the Physiological Laboratory* in 1873 marked a turning point. Sanderson (1829–1905), a professor of physiology at the University of London who had studied with Bernard in Paris, offered the first detailed laboratory manual in English of the methods of French and German physiologists. Not only did the *Handbook* have an immediate impact on British physiology; it also had an immense impact on the British antivivisection movement. Here in print, with copious illustrations, was an animal lover's worst nightmare.

The combination of Sanderson's book, a renewed attack on Moritz Schiff, and a vivisection demonstration at the annual meeting of the British Medical Association (BMA) in 1874, once more brought the issue of experiments on animals before the British public. This time the flurry of articles and editorials escalated into a blizzard. Frances Power Cobbe concluded that new legislation was the only way to restrict the practice of animal experimentation (commonly referred to as vivisection, although not all of it was) in Britain. Richard Martin's Act of 1822, even as subsequently amended, was not strong enough. Cobbe urged the RSPCA to take action. But the RSPCA was a large and diverse group and failed to agree on how to proceed. Finally, she took matters into her own hands. In the summer of 1875, Cobbe's agitation led to the appointment of a Royal Commission to investigate the practice of vivisection.

The RSPCA remained cautious on the issue of vivisection, and in 1875, while the Royal Commission deliberated, Cobbe founded the Victoria Street Society for the Protection of Animals from Vivisection. The Victoria Street Society, whose logo was a faithful, sad-eyed dog, marked the beginning of the modern antivivisection movement. Cobbe used her many connections and friends to gain patronage and prestige for her new society, and the poets Alfred Tennyson and Robert Browning were among its early supporters. The Society sponsored public meetings and petitions in an effort to lobby the Royal Commission. The commission indeed recommended legislation to regulate vivisection, but the bill introduced by Cobbe's allies in the spring of 1876 restricted vivisection far beyond the commission's report. This bill allowed vivisection only when the knowledge to be gained was of clear medical benefit to humans, and dogs and cats were entirely exempted from scientific use. The scientific community, newly self-conscious as a result of the commission, lobbied heavily to modify this bill, and even though Queen Victoria herself barraged the government ministers with antivivisection memos, a compromise view prevailed. The bill was modified, much in favor of the scientists. What is significant, however, is that the Cruelty to Animals Act of 1876 was the first attempt by a national government to regulate animal experimentation. It remained in force until replaced by new legislation in 1986. The 1876 act required the Home Secretary to license and register all experimenters, and to register and subject to inspection all places of experiment. Experiments for teaching and demonstration, as well as experiments without anesthesia on dogs, cats, horses, mules, and donkeys required special certification. Applications for licenses and certification had to be endorsed by a recognized scientific authority, such as the Royal College of Physicians of London. Fines were established, and inspectors appointed.

Cobbe and other antivivisectionists viewed the 1876 Act as a sell-out to the scientists, who were given the final word on the validity and endorsement of research. From this point on, the scientists and the antivivisectionists would become increasingly polarized, with animal protection groups such as the RSPCA occupying an uneasy middle ground. In 1876, the antivivisection movement had barely begun, and following the Act it quickly became radicalized. In 1878, the Victoria Street Society declared it would fight for the total abolition of vivisection, severing its ties with more moderate reformers. Cobbe continued her attempts to introduce new bills into Parliament, with little success. British physiologists continued to reorganize their enterprise on the continental model. In retrospect, 1876 was the last opportunity for compromise between scientists and antivivisectionists. In 1898, the Victoria Street Society ousted Cobbe as its president and changed its policy from the abolition of vivisection to the more moderate goals of restriction and regulation. Cobbe founded the Union for the Abolition of Vivisection, but she had plainly been relegated to the fringes of the movement. Meanwhile, the new science of bacteriology, which relied heavily on animal experimentation, began to give human and veterinary medicine the power to heal (see Chapter 5).

The Special Feelings of Women

Although men usually occupied the highest positions in the antivivisection movement, at least half of those active against vivisection were women. In a period when most women—especially middle- and upper-class women—had little opportunity to enter public life, this is a remarkable figure. Why were so many of these women involved? The notion of "separate spheres" had served to relegate middle- and upper-class Victorian women to the household more than in the past, differentiating them sharply from the industrial working class. Frances Power Cobbe urged the involvement of these women in the campaign against vivisection with a combination of traditional and feminist ideology. Women, she said, are naturally more sensitive and spiritual, and therefore more attuned to the plight of animals. Yet women's higher moral character also made it essential that they enter public life and not just operate from the home. Cobbe persuaded many moderate women, who were put off by feminism and the suffrage movement (in which Cobbe was also involved), by showing them that they need not sacrifice their "womanliness" to be active. In this she effectively countered popular arguments used by antisuffrage men and women. These arguments also persuaded middle-class American women to become active in antivivisection.

For feminists such as Cobbe herself, antivivisection offered another outlet for their energies, giving them valuable experience. But nonfeminist single women, seeking a charitable cause, also found antivivisection attractive. Other women had differing motives. In France, Claude Bernard's estranged wife and their daughters rescued experimental animals, especially dogs, and also cared for stray dogs from the streets of Paris. A story, probably fictional, tells of the search by Bernard's daughter for the pet dog of a friend, only to find it on the vivisecting table of her father. Mme. Bernard and her daughters spent much of their fortune caring for sick and stray dogs and cats, and Mme. Bernard was a long-time member of the French antivivisection society.

The physicians Elizabeth Blackwell (1821–1910; the first woman to graduate from an American medical school) and Anna Kingsford (1846–1888) saw parallels between the rise of surgery on women, especially gynecological surgery, and vivisection. While in medical school, Blackwell and Kingsford had witnessed how public hospitals used poor women as teaching tools for student doctors. At a time when showing the ankles was considered immodest, these two women were horrified to see other women's bodies exposed to the cynical gaze of male medical students. Because, as we have seen, the poor were thought to be less sensitive to pain, surgical operations, even in the 1880s, were performed without anesthesia. Anna Kingsford explicitly compared the poor, women, and animals—all considered fit subjects for painful experiments. Many physicians, both male and female, had doubts about the rise in surgery, such as removal of the ovaries or clitoris, as a solution to "women's problems," which could include "hysteria." Kingsford and Blackwell, however, went on to connect this to vivisection, and Blackwell referred to the removal of ovaries as "castration." Blackwell saw only the thinnest of lines between animal vivisection and experimentation on poor women. Kingsford was more extreme, offering herself as a replacement for experimental animals, and cursing Claude Bernard and other experimenters. The patrician Cobbe felt that Kingsford's behavior only reinforced the notion of the overly emotional woman.

Fictional works such as the popular *Black Beauty* (1877), a biography of a horse written by the English Quaker Anna Sewell, dramatized animal suffering. In the 1890s, two best-selling works of fiction portrayed the scientist in widely differing ways, betraying the deeply conflicted feelings of the public over scientific advances. Arthur Conan Doyle, a physician, introduced Sherlock Holmes in 1887, white-coated, in a laboratory. Holmes did not hesitate to prick himself in the finger to obtain a blood sample, and he looked at humanity with a cold and clinical eye. H. G. Wells, who had studied science at the University of London and authored a biology textbook, presented the antivivisectionist's nightmare in *The Island of Doctor Moreau* (1896). Although in the book Moreau was ostracized from mainstream science for his horrific vivisections, the implication nonetheless remained that science, unrestrained, was capable of still greater horror, unless nature herself rebelled.

In his 1875 defense of animal experimentation, the American physiologist John Call Dalton (1825–1889) could point to increased knowledge of physiology but few practical medical advances. The antivivisection movement based its arguments on the notion of needless pain, for scientists had not demonstrated to the public that their experiments were worthwhile. As we shall see in the next chapter, twenty years later this picture had changed greatly, and the advent of microbiology promised to eradicate human disease for good, with the help of the bodies of millions of animals.

5 The Microbe Hunters

In 1878, the year Claude Bernard died, a Frenchman coined the word *microbe* to denote a microscopic infectious agent. It gradually became clear that "microbe" encompassed a great variety of organisms, including animals such as bacteria and protozoa and plants such as fungi and yeast, each of which caused infection in different ways. In addition, the infectious agents for diseases such as smallpox and rabies could not be found and were presumed to be too small to be viewed under conventional optical microscopes. We now know that these diseases are caused by viruses, which are much smaller than other disease-causing microbes. The adoption by scientists and physicians of the germ theory of disease, which stated that contagious diseases were transmitted by means of the action of microbes, opened up a new arena for research that promised immediate benefits for humans. In contrast, although Bernard and other physiologists promised that their research on animals would lead to the improvement of human health, they placed these goals far in the future. Physiology and microbiology differed in important ways, but each discipline depended on human and especially animal bodies for experimental proof.

By the time Frances Cobbe was ousted as president of the Victoria Street Society in 1898, the promise of science to improve human health was no longer a vague goal, but a reality. New microbes were being discovered almost daily, and a new vocabulary of vaccines, immunizations, and antitoxins entered everyday life. Antisepsis and asepsis, together with anesthesia, revolutionized surgery, making complex operations possible and elevating surgeons from doctors of last resort to the elite of the medical profession. Dozens of scientists contributed to these developments, but two men were largely responsible for the development of the germ theory of disease in the 1870s and its subsequent public acceptance: the French chemist Louis Pasteur (1822–1895) and the German physician Robert Koch (1843–1910). By using thousands of animals and a few human subjects, they developed the theories and techniques that received their fullest expression in the years before World War I in the Frankfurt laboratory of Paul Ehrlich (1854–1915). With Ehrlich's work, the laboratory joined decisively with the clinic in the work of preventing and treating disease, and "scientific medicine" became a reality. Yet, as is always true of science, this process was neither straightforward nor inevitable, and the germ theory was not without its critics. Despite the heroic stature subsequently granted the founders of microbiology by press and public, ambiguities and ethical lapses were also part of the process.

Pasteur and Microbes

One hundred years after his death, Pasteur is undoubtedly still the most famous scientist in France. Every French village has a rue Pasteur, and when, a few years ago, an American historian dared to suggest that Pasteur might have been motivated by ambition and competitiveness, a national furor erupted. Yet Pasteur's early career made him a somewhat unlikely candidate for the founder of bacteriology. The son of a provincial tanner, Pasteur trained as a chemist, not a physician. His early studies of the optical qualities of the crystals of certain organic acids led to the interesting discovery that some crystals were not symmetrical, and that indeed they exhibited left- and right-handed versions. Pasteur found that only the molecules of living substances exhibited this asymmetry. Thus he came to the study of biological problems from the study of chemical problems; but he came convinced, like Bernard, that biology was somehow different from the physical sciences.

In the 1850s, Pasteur had an appointment at the University of Lille, in northern France, where the distillation of juice from sugar beets into alcohol was an important local industry. Pasteur began to study the process of fermentation, by which alcohol was produced from the juice. Fermentation helped to turn grapes into wine, flour into bread, milk into cheese. But no one knew how it worked. Fermentation was related to putrefaction: rotting was the next step in the process, and it often occurred despite efforts to prevent it. By means of chemical analysis and microscopic investigation of fermenting beet juice, Pasteur concluded that fermentation was not an inorganic chemical process but was caused by a living organism, microscopic in size, called yeast. Pasteur's fellow chemists were not happy with this mingling of inexact biology with rational, quantitative chemistry. While physiologists were endeavoring to take the messy uncertainty out of biology by reducing it as much as possible to physics and chemistry, the chemist Pasteur put uncertainty back in. In 1859 the Paris Academy of Sciences awarded Pasteur its prize in experimental physiology for his work on fermentation. The chair of the prize committee was Claude Bernard.

The implications of Pasteur's explanation of fermentation for the theory of disease were not immediately evident, least of all to Pasteur. In Paris in the 1860s, Pasteur took on the problem of spontaneous generation, the age-old question of whether life could come from non-life. Contemporary theorists argued that a vital principle in nature could spontaneously organize inorganic substances into living ones. Pasteur argued that only living microorganisms, which were in the air, could produce more organisms. By means of a series of experiments involving infusions in sealed and unsealed flasks, Pasteur demonstrated that the flasks left open to the air generated microorganisms, while those heated and sealed from the air did not. In fact, there was a certain amount of luck in Pasteur's success, since as he later found, some microorganisms could withstand heat, and even seemingly clean flasks could be contaminated.

In the mid-1860s, Pasteur was asked to investigate an epidemic of a disease known as "pébrine," which attacked silkworms (the caterpillar of a moth). Silk was a major industry in certain parts of France, so that rescuing the silkworm was of national importance. But Pasteur did not yet have a clear idea that microorganisms could cause disease. By the example of fermentation, a microorganism might affect chemical changes in its host, but not kill it. Current medical debates on disease causation centered on the idea of a "miasm" or atmospheric condition that could bring on certain diseases. Public health measures, including cleaning streets, purifying the water supply, and providing sanitary waste disposal were being implemented in Britain and elsewhere to stem the tide of epidemic diseases such as cholera and tuberculosis, which flourished in the damp and dirty conditions of urban industrial slums. Microorganisms played little role in this picture.

Pasteur soon found microorganisms in silkworms. But what did they do? At first Pasteur believed the "corpuscles," as he called them, were only the symptoms of pébrine, and that they only came into view under the microscope when the disease was advanced. But he gradually came to recognize that the microorganism indeed caused the illness, and that it did not travel through the air but was transmitted through the moths' eggs. At the same time, he identified another silkworm disease known as "flâcherie" and concluded it too was caused by a microorganism, in this case one that flourished on damp mulberry leaves, the preferred food of silkworms. Hygienic measures, such as keeping the silkworms' beds clean and dry, prevented the disease. Thus in both cases, he had shown that a specific microorganism caused a specific disease.

Next he developed the process of pasteurization, which killed certain microorganisms that caused spoilage by applying heat for a specified period of time. This process was first used for wine—another major French industry—and then also for beer and later, milk. In the 1860s, the British surgeon Joseph Lister had applied Pasteur's ideas about microbes to the phenomenon of postsurgical infection, and introduced antiseptics to kill microbes on tools and in the air, which made surgery much less risky. Aseptic practice—preventing infection with sterile techniques and tools rather than with chemical antiseptics—was even more effective. Pasteur himself promoted sterile techniques and experimented on dogs with different kinds of wound dressings.

In 1865, Pasteur and Bernard had been members of a commission to study cholera, which was about to reach epidemic proportions in Paris. Pasteur was convinced that a microorganism caused the disease, but he had no idea how it acted, and he believed it was carried through the air. He analyzed the air of a cholera ward, as well as dust from the floor and bedding, but failed to find an infectious agent. Bernard took blood samples, but analyzed them chemically, and did not look for microbes. A dozen years later, when Pasteur began to look into the cattle disease, anthrax, his experience of pébrine and flâcherie led him to recognize that the air may not be the only vehicle of infection.

In the 1850s, a French physician named Davaine had noticed a rod-shaped microbe in the blood of anthrax victims. A decade later, following Pasteur's fermentation experiments, Davaine injected sheep's blood that he believed to be infected with anthrax microbes into rabbits, which soon died. But others pointed out that the rabbits may not have died from anthrax: cow's blood with no evidence of the microbe was equally fatal to rabbits. In 1877, the same year Pasteur began to work on anthrax, a rural German physician named Robert Koch grew the anthrax bacteria outside of the body, using the aqueous humor of an ox's eye as the medium. Under the microscope, he could see the characteristic rods grow and multiply.

Anthrax is the biggest disease-causing bacteria, and the easiest to see and distinguish from other microorganisms. In addition, Koch's new staining techniques made the anthrax rods even more evident. In 1877, Pasteur and Koch each published findings on the etiology (causes or origins) of anthrax, describing its life cycle, how it got into the blood, and how it caused disease. This was the first widely accepted demonstration of a bacterial cause of disease. In a paper on wound infections published in the next year, Koch introduced his famous "postulates," which governed bacteriological research:

1. the microorganism in question should be found in each case of the disease, that is in each experimental animal
2. the microorganism should not be found in cases of other diseases (specificity)
3. the specific organism should be isolated

4. the organism should be cultured, that is, grown outside the body in a sterile medium
5. the organism should, when inoculated into a well animal, produce the same disease
6. the same specific organism should then be recovered from the diseased animal

Koch's postulates emphasized the cultivation of the suspected microbe outside the human or animal body, in the controlled environment of a sterile artificial medium. Koch himself increasingly used a variety of gelatin. Animals were used to retrieve and to test the specificity of the microorganism: they were receptacles for the microorganism just as the dishes of gelatin were. In developing vaccines for anthrax, chicken cholera, and swine fever between 1878 and 1883, Pasteur followed Koch's techniques, with great success. In each of these cases, Pasteur developed a vaccine by weakening the virulence of the microbe, whether by heat, chemicals, or prolonged exposure to the air. The principle of a vaccine, he believed, must follow the model of Edward Jenner's smallpox vaccine: a weaker variety of the disease in question could provoke immunity.

The microbe itself was cultivated in a sterile medium. But in these studies, Pasteur and his assistant, the physician Émile Roux (1853–1933), discovered that the virulence of a microbe could also be altered—either increased or decreased—by passing it through living animal bodies. Pasteur passed attenuated (weakened) cultures of chicken cholera through a series of chickens to increase its virulence. Koch, who had also noticed this phenomenon, thought that the repeated passages acted to purify the microbe. But Pasteur gradually came to realize that the microbe itself changed—what is now called mutation.

The Rabies Vaccine

Rabies is a nasty animal disease, caused by a virus, which humans occasionally contract, generally by being bitten by a rabid animal. It attacks the central nervous system and sometimes causes an inability to swallow—thus its alternative name of "hydrophobia," or fear of water. Although, in the nineteenth century, rabies was uncommon both in animals and in humans, it excited particular dread for several reasons: its long incubation period (several months in humans) meant an extended period of anxiety after possible exposure; its symptoms were horrific, painful, and invariably fatal; and it was most commonly carried by the domestic dog, which in this period was consolidating its status as the companion animal of choice.

In economic terms, an anthrax vaccine was more valuable than one for rabies. Far more economically valuable animals such as cattle died of anthrax. Very few people died each year of rabies, and even if one were bitten by a mad dog, rabies did not inevitably result. But the high public profile of rabies meant that the fame Pasteur gained for his rabies vaccine far surpassed what he had enjoyed before. The dramatic story of its first human trial, on the boy Joseph Meister, added to Pasteur's public acclaim.

Pasteur and Roux began their quest for a rabies vaccine in 1880. Pasteur could not find the rabies microbe since, as a virus, it was not visible through his microscope, but he remained convinced that the cause of rabies was a microorganism, and that the techniques he had used to develop vaccines for other animal diseases would also be effective with rabies. Pasteur at first believed that immunity could only be conveyed by a living but weakened microbe, which would consume the nutritional supplies of the host organism and leave nothing for the more virulent microbes that followed. Later he concluded that the microbe itself released a chemical toxin and that immunization was therefore a chemical process. If this seems like a surprising slip for Pasteur the chemist, it also serves to remind us of the immense uncertainty surrounding microbes and their action in this period. Pasteur thought he understood what caused immunity, but he did not know how the body produced immunity. While today, "germs cause disease" seems a statement too obvious to need explanation, barely a century ago every word in that statement was questionable: What was a germ? What was disease? How was disease caused?

Building on their previous experience, Pasteur and Roux employed animal bodies, rather than sterile gelatin, as their cultural medium in their quest for a rabies vaccine. They had already noted that passing a disease through a number of individuals of the same species could markedly affect its virulence. After a certain number of passages the microbe would become "fixed" or stable. In addition, a French veterinarian reported in 1879 that when rabies was transmitted from dogs to rabbits, the month-long incubation period for the disease in dogs dropped to about eighteen days in rabbits. Pasteur found that additional passages through rabbits reduced the period even further, to six or eight days. In addition, rabbits and guinea pigs, which he also used, were more docile than rabid dogs.

Pasteur's decision to use animals as his test tubes had momentous consequences for the future use of animals in research. Thereafter, bacteriological and immunological research became inextricably linked to the use of animals as culture media. As the scale of research grew, so too did the numbers of animals. Pasteur used hundreds of animals to develop the rabies vaccine; between 1882 and 1885, for example, he passed the virus through ninety sets of rabbits.

Paul Ehrlich, twenty years later, used thousands of mice to develop Salvarsan, as we shall see below. The development of the polio vaccine in the 1950s, discussed in Chapter 6, required that millions of monkeys be sacrificed. Physiological research, in contrast, employed a fraction of these numbers.

Pasteur began his rabies research as he had his other disease research: he injected several animals with different substances—saliva, blood, or tissue from rabid dogs—and then waited to see what would happen. He discovered that nervous tissue, and especially the brain, was the ideal medium for cultivating the rabies microbe, and he claimed that injecting rabid brain matter into animal brains invariably caused the disease. He also used monkeys, which were rarely used at the time for research, finding that while passage of rabies through rabbits made the virus more virulent to dogs and presumably humans, passage through monkeys made it less virulent. A sufficient number of passages through monkeys would weaken the microbe enough, he surmised, to make an effective vaccine, and in the summer of 1884 he proclaimed that he tested such a vaccine on dogs and that it worked.

Pasteur worked on several methods of attenuation besides the monkey method. One method involved drying the spinal cords of affected rabbits. To immunize the dogs, he injected them with portions of spinal cords of gradually increasing virulence. It should be noted, however, that the nature of immunity and its acquisition was so imperfectly understood that at first he injected them with cords of gradually *decreasing* virulence, which would be more likely to induce the disease.

In 1885, Pasteur was on the verge of success. He continued his experiments in inducing immunity in dogs, and began a series of experiments to test if the method of immunization would work even after the victim had been exposed to rabies, because of its long incubation period. If it was possible to inoculate a victim after exposure had occurred, and prevent the disease from manifesting itself, then Pasteur's vaccine could also act as a cure for a most dreaded disease. According to historian Gerald Geison, who closely examined Pasteur's laboratory notebooks, the scientist attempted to cure two humans by his method of inoculation in the spring of 1885. In both cases the required series of injections were not completed, and the results were inconclusive. Up to that point, experiments on exposed dogs were also inconclusive.

On a hot July day in 1885, nine-year-old Joseph Meister, the son of a baker in Alsace, in eastern France, sat trembling in the office of the famous Doctor Pasteur. He was weary from the long train trip from Alsace to Paris, and the many dog bites on his hands, arms, and legs hurt horribly. He could still hear the yelps of the dog and feel its hot foul breath on him, and hear his own

The Rabies Virus

An American magazine, *Harper's Weekly*, reported on Pasteur's experiments on 21 June 1884, describing his procedures and the scene in his laboratory. The writer commented, "The experiments, cruel as they may appear at first sight, are made in the interest of humanity, and M. Pasteur is careful not to inflict needless suffering on the dumb creatures which he subjects to the operation," noting that he chloroformed the dogs before injecting the rabies virus into their brains. The article emphasized dogs, barely mentioning rabbits, guinea pigs, and monkeys.

■ *Harper's Weekly*, 28:1435 (1884): 395, illus. on 392.

screams. The local physician had cauterized some of the bites with phenic acid, a standard treatment, which only made them hurt more. Monsieur Vone the grocer, the owner of the dog, had come with them on the train. The dog had also bitten him, but not as severely. He had killed the dog, who was obviously mad. Everyone in the village said that Pasteur was Joseph's only hope, or he would die a terrible death.

Pasteur brought two physicians in to see Joseph, and they agreed that the number and severity of the bites indicated that the boy would almost certainly develop rabies. Pasteur made a decision. In the paper of October 1885 in which he described the case, he wrote, "The death of this child appearing to be inevitable, I decided, not without lively and sore anxiety, as may well be believed, to try upon Joseph Meister the method which I had found constantly successful with dogs."* But according to the evidence of his notebooks, Pasteur had not in fact completed the relevant experiments on dogs. Nonetheless, he decided to go ahead and treat Joseph. Over a period of ten days, he injected the child with increasingly virulent preparations of dried rabbit spinal cord, ending on July 16 with the most virulent. Joseph did not come down with rabies—neither with the natural version he had acquired, nor with the injected rabbit version, which would have manifested itself much more quickly. By mid-August, Joseph was declared cured, and Pasteur was a national hero.

As his paper was being published in October, Pasteur was treating another young rabies victim, a shepherd named Jean-Baptiste Jupille. He also did not contract rabies, and by the end of 1885, a portion of Pasteur's laboratory had become a clinic for dog-bite victims. In 1889, the Institut Pasteur opened, the first publicly endowed research laboratory in France. Both Meister and Jupille later became guards at the Institut, and Meister was shot defending Pasteur's laboratory during the German occupation of Paris in World War II.

Was Pasteur's treatment of these boys unethical? By modern standards, absolutely. To perform human trials without adequate demonstrations on animals is a violation of modern clinical and experimental ethics, and if Pasteur were in a modern laboratory, answerable to institutional ethics committees, he would not have been allowed to go through with his treatment. Even his assistant, Émile Roux, believed that Pasteur had inadequate animal evidence on which to base his conclusions, and he refused to inject Meister or the others (since Pasteur was not a physician, he could not perform the injections himself). On the other hand, Joseph Meister seemed to face certain death. We now know that not everyone exposed to rabies develops the disease, but at the time

* Louis Pasteur, "Prevention of Rabies" (1885), in Elie Metchnikoff, *The Founders of Modern Medicine: Pasteur, Koch, Lister* (New York: Walden, 1939), 383.

opinion was unanimous that the boy would die without treatment. In addition, neither Meister nor his mother, nor Jupille could be said to have been adequately informed, by modern standards, of the possible results of their treatment. The ethical dilemma of Joseph Meister's case, then, is on several levels. Did Pasteur believe he was acting ethically, or was he motivated by ambition? Or are these motives mutually exclusive? Was Roux correct to assert that no human trials were possible, given the current state of knowledge? Can we apply modern ethical standards to activities in the past? With present restrictions on research, could Pasteur have made his discoveries today? These questions are not easily resolved.

Koch and the Antivaccinators

At the time of his 1877 paper on anthrax, Robert Koch was an obscure German physician in what is now Poland. Working alone in a back room of his house, Koch developed experimental techniques critical to the development of bacteriology. He used the microscope extensively, but one of the problems of looking at bacteria under the microscope, especially ones smaller than the relatively large anthrax rods, was their lack of contrast against the tissue background. Koch developed staining techniques, methods of fixing tissue specimens, preparing slides, and especially, photographing specimens. His published papers were among the first to employ photographs. To skeptics of the germ theory, these photographs were especially convincing, showing that different microorganisms looked different and supporting the theory that specific microbes caused specific diseases.

As we have seen, Koch also developed in vitro culture techniques. He grew the disease organism in an artificial environment of a sterile medium. By this means, Koch could demonstrate how a microorganism reproduced, and could then induce disease in a well animal—which could vary from a cow to a rat—by injecting the microorganism. Natural infections, especially wound infections, which he studied in the late 1870s, were far too variable to afford reliable proof. By isolating a single disease-causing set of microbes, Koch eliminated the chance infections that could destroy an experiment.

In 1880, the German government gave Koch a laboratory in Berlin. He left his medical practice for research. In contrast, Pasteur was not given his own research institute until 1889, at the end of his career. The German government's support of science meant that until World War I, it led the world in microbiology. Koch inaugurated his laboratory by tackling tuberculosis, one of the most baffling and deadly diseases of the era. Tuberculosis was known primarily as a disease of the lungs, in which resided its most evident symptoms, and nineteenth-century literature is filled with wasting, feverish tuberculosis victims, whose bloody coughs were a death sentence. Yet tuberculosis could also reside in a seemingly well human or animal for many years without provoking symptoms. There was no known cure for tuberculosis, and much debate about its causes. Its trail of contagion was difficult to trace, since one could contract the disease by contact with someone who displayed no symptoms.

Koch succeeded in isolating the tuberculosis bacillus by staining tissue from tuberculosis patients. He addressed the Berlin Physiological Society in March 1882 on tuberculosis, and the paper, published soon after, created a sensation. Koch went through his postulates and described several experiments on animals. He demonstrated that, as in anthrax, the tuberculosis bacillus had a spore stage during which it could be transported through the air and remain viable in a dry atmosphere that would normally kill it.

Although Koch showed that tuberculosis could be carried by milk, he did not believe that cows could transmit bovine tuberculosis to humans via their milk. Around 1900, the American bacteriologist Theobald Smith (1859–1934) demonstrated that children, especially, could indeed be infected with bovine tuberculosis by drinking contaminated cow's milk, and his results led to the widespread pasteurization of milk and more sanitary farming conditions, which helped check bovine tuberculosis. But attempts to develop a tuberculosis vaccine failed. Koch announced in 1890 that he had developed a substance called "tuberculin," which cured tuberculosis. He did not reveal its composition, but claimed its effectiveness in animal trials. But tuberculin, a preparation of cultures of tuberculosis bacilli, proved to be ineffective and dangerous. The French physicians Léon Charles Albert Calmette (1863–1933) and Camille Guérin (1872–1961) developed a live bacillus tuberculosis vaccine in 1906, which came into general use in the 1920s. However, the BCG vaccine, as it is known, carries some risk of infection, and those vaccinated then test positive for tuberculosis, making detection of the natural disease difficult. In many countries, including the United States, it is little used.

Between 1880 and 1900, the validity of the germ theory of disease became increasingly evident, as the microbial causes of several diseases were isolated. Yet scientists disagreed about practically every important issue, including how infection occurred, how the body resisted infection, what immunity was, and how vaccines worked. Pasteur's rabies cure gained much public attention, but so too did Koch's failure to develop a tuberculosis vaccine or cure. The discovery of the malaria parasite in 1880 enabled physicians to use the centuries-old medication of quinine only in malaria and not in other fevers, and tests were developed to detect the presence of other disease-causing microbes, making diagnosis more accurate. Yet the more microbes were discovered, the more variable seemed disease. Nonetheless, laboratory techniques, including animal use, continued to be developed and refined, and microbiology firmly established itself as a discipline between the laboratory and the clinic.

A major success was the development of diphtheria and tetanus antitoxins in Koch's laboratory by the Japanese Shibasaburo Kitasato (1852–1931) and the German Emil von Behring (1854–1917). Extending the work of Émile Roux on diphtheria, they found that diphtheria and tetanus, two very different diseases, worked by essentially the same mechanism: the bacteria multiplied in a local area, and produced a poison or toxin, which then circulated in the body—in diphtheria, via the bloodstream; in tetanus, via the nerves. They showed that they could induce tetanus in a well animal by injecting it with the bacteria-free (and cell-free) serum of a sick animal, usually a guinea pig: clearly the toxin alone caused the disease symptoms. Following Pasteur's principles of inducing immunity, Kitasato and Behring then showed that animals injected with sublethal amounts of disease toxin became immune, and they could after a time survive a lethal dose of toxin. Moreover, other animals injected with the serum of immune animals gained their immunity. But antitoxin was a therapy, not a vaccine. It could not control the spread of disease from person to person, and the injected antibodies did not last long in the body.

Many people, lay people and physicians, remained unconvinced of the value of the germ theory. The miasma theory pointed to dirt, overcrowding, bad air, and unclean water as sources of disease, without reference to the action of microbes. Bad smells and dirt could themselves cause disease. Public health advocates noted that improved sanitation greatly reduced the incidence of diseases such as cholera. The inconsistent behavior of microbes, and the failure to isolate microbial causes for all diseases (viruses continued to be poorly understood) discredited the germ theory in some eyes. In 1884 Koch isolated the bacillus that causes human cholera, but could not then induce the disease in animals. Anticontagionists argued that cholera, therefore, was not contagious. In the 1892 cholera epidemic, scientists conclusively showed the presence of the cholera microbe in the water supplies of cholera-ridden cities, but some still were not convinced. The German health reformer Max Pettenkofer swallowed a glass full of microbe-laden water and did not fall ill; he proclaimed triumphantly that the germ theory was wrong. In fact, he was amazingly lucky. The Russian composer Peter Ilyich Tchaikovsky did the same thing, although not on purpose, and died of cholera.

Prominent opponents of the germ theory, particularly in Britain, were also strong antivivisectionists. A well-known physician, Sir Benjamin Ward Richardson (1828–1896), argued that a clean environment, clean bodies, and clean minds would eliminate disease, but experimentation on animals would not. Frances Power Cobbe would certainly have agreed with these sentiments, although she also thought that undue attention to one's health was mere selfishness. The language of purity and impurity in the public health movement also condemned the introduction of foreign substances such as vaccines into the body. How could injecting diseased tissue—particularly diseased animal tissue—make you well?

The playwright George Bernard Shaw (1856–1950) offered a witty but scathing attack on the germ theory, vaccines, vivisection, and the medical profession as a whole in his 1913 play *The Doctor's Dilemma*. In a lengthy introduction to the play, Shaw, a prominent Socialist and antivivisectionist, skillfully exposed the fears of the populace and the pretensions of scientific medicine. The poor and ill, he said, do not need medicine but better food and housing, and the new scientific medicine was designed not to keep people in health but to cure diseases. Having lived through the vogue for public health, followed by the vogue for germs, Shaw was skeptical about new theories: "I have heard doctors affirm and deny almost every possible proposition as to disease and treatment."* How can doctors say that all diseases are caused by germs when they have not seen the germs of many of them? The germ of smallpox, after all, had still not been seen. Was not bacteriology mere belief and superstition then? If science was doubtful, common sense and cleanliness should win the day.

With the anticontagionists, Shaw believed that immunizations were dangerous and misguided. Koch's tuberculin fiasco showed that doctors simply followed the latest trend with little knowledge of the actual causes of disease. Shaw characterized immunizations as a plot to make money for physicians, claiming that the majority of people could not afford them. Having demonstrated that physicians are slavish followers of fashion, he characterized their support of vivisection as simply another facet of this. The right to knowledge, he argued, does not include the right to cause pain, and indeed doctors above all should not cause pain—an argument also used by Benjamin Ward Richardson, who developed several new kinds of anesthesia to minimize animal and human pain.

Shaw also tweaked the upper-class antivivisectionists who criticized scientists but did not hesitate to hunt, eat meat, or wear furs. His conclusion, that medicine should be socialized, probably did not sit well with some who otherwise agreed with him, but Shaw expressed widely shared doubts and confusion

*George Bernard Shaw, *The Doctor's Dilemma* (1913, reprint, Harmondsworth: Penguin, 1954), 29.

about the new scientific medicine. Was medicine about to abandon the bed-side for the laboratory, or could the two coexist? Nearly ninety years later, this is still a relevant question. On the other hand, the promise of finding cures for the ills of humankind might justify any measures.

Ehrlich and Salvarsan

Shaw's skepticism about the capability of scientific medicine to cure disease was being challenged even as he wrote. In 1907, Paul Ehrlich patented a substance he called 606, later known as Salvarsan. Number 606, he reported, could cure syphilis. It was called 606 because it was the 606th compound Ehrlich had tried. Only in 1905 had two German scientists identified the corkscrew-shaped microorganism of the variety known as spirochetes whose scientific name became *Treponema pallidum*. Several other microorganisms had been identified in the previous two decades as the cause of syphilis, but the treponemes were found consistently at all stages of syphilis and in all affected organs. In the next year, another German, August von Wassermann (1866–1925), gave his name to the Wassermann test, which accurately diagnosed syphilis on the basis of the presence of antibodies in the blood.

There is much debate about the origins of syphilis in Europe. Some have argued that Columbus brought it back from the New World; others, that it was already in Europe. In any case, around 1495 a virulent and often fatal form of syphilis began to spread in Europe, and it remained the most serious sexu-ally transmitted disease until AIDS. Syphilis usually begins as a sore at the site of infection, but then spreads to the body's internal organs over a period of years, often involving the heart or nervous system. Because its symptoms can vary, it is difficult to identify on that basis, and it was often confused with gon-orrhea, another venereal disease with somewhat less serious effects.

Ehrlich had developed several chemical stains and dyes for Robert Koch, and in the 1890s, he worked at Koch's research institute. His guiding principle from the outset was that biological activities are determined by specific chem-ical affinities and are quantitatively measurable. This principle opened the possibility of chemical cures for disease. In the 1890s, Ehrlich worked on the production of immunity in animals, not with microbes, but with toxic plant proteins such as ricin, developed from the castor plant. He found that by ad-ministering small doses of the poison and gradually increasing them, the ani-mal's body would produce specific antibodies against the poison. This is one principle by which disease immunities are developed, but in this case no dis-eases were involved. Ehrlich noted that this immunity could be transferred to an infant via its mother's milk.

At Koch's Institute for Infectious Diseases, Ehrlich then looked at bacterial toxins in the same way as he looked at plant toxins, and played a role in the development of diphtheria antitoxin, one of the great successes of the era. In 1899, he moved to a new institute built for him in Frankfurt, where he remained until his death. Germany's many state-run research institutes were critical to the growth of bacteriological science, but also had another effect. Any therapeutic agents discovered, such as antitoxins, remained in the hands of the state rather than of a private company.

Ehrlich continued to seek specific chemical affinities between diseases and certain chemicals. His research program intensified between 1906 and 1910, as he attempted to discover synthetic chemicals that acted specifically on pathogenic microorganisms. Could he find affinities between specific chemicals and specific organisms, and nothing else in the body? Could he find an "internal disinfectant" that acted as well as Lister's external ones? Ehrlich used the images of "magic bullets" and "poisoned arrows," which zeroed in on a "parasite" and barraged it with chemicals. He tested his "magic bullets" on isolated disease organisms, first in vitro, but then on live animals, usually mice. He believed that only live animals could give a valid assessment of a drug's potency and safety.

Ehrlich employed industrial chemists from the dye industry to make compounds, and tried several hundred. He found that the dye "trypan red" appeared to cure mice infected with *trypanosoma,* protozoans that caused the tropical disease known as sleeping sickness. But the treated animals soon developed resistance to trypan red, and Ehrlich turned to compounds of arsenic. He and his chemists synthesized hundreds of compounds to find the best combination of the least toxicity for the most effectiveness. Compound number 606 seemed especially promising, and Ehrlich patented it in 1907. It killed all the syphilis spirochetes in the test mice without apparent harm to the animal.

As Pasteur was inundated with requests for his rabies cure, so now Ehrlich was overwhelmed with requests for the new syphilis cure. However, he continued to test the drug on animals, and conducted clinical trials on humans in the summer and autumn of 1910 before allowing the drug to be distributed commercially under the name of Salvarsan. The German government, who sponsored Ehrlich's laboratory, held the patent on Salvarsan and closely guarded the secret of its composition. During World War I, American scientists succeeded in duplicating the compound and sold it for much-reduced prices.

Salvarsan was not a perfect drug, despite the years of tests. Syphilis followed a complex pathology of several stages, and Salvarsan was effective only in its early stages. Ehrlich's recommendation for one or two large doses of the drug, rather than a series of smaller doses, led to dramatic cures but also to dramatic and sometimes fatal side effects, caused by the toxicity of the drug itself or by allergic reactions to the massive die-off of spirochetes. Close scientific supervision of the drug's complex manufacturing process was also necessary. Ehrlich found a less toxic compound, number 914, a few years later, which was dubbed Neosalvarsan and which became the most-used cure for syphilis until the development of penicillin in the 1940s.

Continued use of the drug by physicians established a lower effective dose and a longer period of therapy. Doctors correlated the clinical progress of Salvarsan and Neosalvarsan with the Wassermann test for syphilis antibodies, providing a model for the new scientific medicine, which depended on the laboratory for every step of the healing process.

Antibiotics and the Decline of Antivivisection

Despite the success of Salvarsan, few other comparable chemical drugs appeared for nearly twenty years after Ehrlich's death in 1915. Bacterial infections remained without remedy. Diphtheria antitoxin, obtained mainly from horses, was a notable success. Antitoxin or serum therapy, as we have seen, was not applicable in all infectious diseases and of limited use as a preventive. But in 1907, Theobald Smith experimented with guinea pigs using a mixture of diphtheria toxin and antitoxin, which appeared to confer long-lasting immunity. By 1913 Emil Behring developed a diphtheria vaccine and successfully tested it on humans. Coupled with the Schick test for diphtheria antibodies (invented by Bela Schick [1877–1967] in 1913), the toxin-antitoxin vaccine, even more than Salvarsan, represented the power of scientific medicine. Diphtheria was a disease of innocent childhood, while syphilis carried the taint of illicit behavior. New York City's health department began to administer the Schick test to all schoolchildren in 1919, and immunized those who tested positive. Rates of infection and mortality plummeted in the 1920s.

Other antitoxins offered mixed results. While tetanus antitoxin was effective as a cure, it sometimes caused fatal allergic reactions. The tetanus toxin was so powerful that a toxin-antitoxin vaccine like that for diphtheria could not be developed. Cholera antitoxin proved to be ineffective, and the decline of cholera had more to do with improvements in hygiene. Attempts at developing other vaccines had similarly mixed results. A killed-bacillus typhoid vaccine was effective, but vaccines for pneumonia and meningitis were not. Microbiology was still an imperfect science, and much remained unknown about the actions of microorganisms. At the Pasteur Institute in Paris, at Koch's and Ehrlich's institutes in Germany, and at universities and research centers around the world, mice, rats, guinea pigs, dogs, cats, horses, and other animals continued to be sacrificed to provide another piece of the puzzle of human and animal disease.

Nonetheless, the successes of microbiology had an enormous public impact. Science, it was clear, held the power to cure all human diseases, if not immediately, then soon. Two popular American books from the 1920s fostered the image of the heroic scientist. In *Arrowsmith* (1925), the well-known novelist Sinclair Lewis (1885–1951) presented the story of Martin Arrowsmith, a midwestern boy who grew up to become a bacteriologist. Although Arrowsmith's devotion to bacteriological research caused him to neglect family and fortune, there was no doubt of his heroic stature. Lewis vividly depicted the lonely and sometimes tedious life of a laboratory scientist, and his McGurk Institute was closely modeled on the real Rockefeller Institute in New York, founded by the oil baron John D. Rockefeller in 1901 as a privately endowed research center devoted to microbiological research. Lewis gained insights into the scientist's life from his friend Paul de Kruif (1890–1972), a former Rockefeller Institute scientist turned journalist. De Kruif's best-selling *Microbe Hunters* (1926) presented the search for microbial agents as a series of highly dramatic stories, written in a racy, colloquial style. While Lewis and de Kruif humanized the scientist, they also fed the public's hope that these ordinary but dedicated men would inevitably find the answers they sought.

In the next decade, as antibacterial drugs finally began to emerge from laboratories, Hollywood joined in the glorification of the scientist. *Arrowsmith* became a successful movie in 1931. In the 1936 blockbuster *The Story of Louis Pasteur,* actor Paul Muni won an Oscar for his convincing performance of the hero battling the forces of ignorance. Paul de Kruif's chapter in *Microbe Hunters* on Walter Reed and yellow fever research was the basis for the film *Yellow Jack* (1938). A few years later, Edward G. Robinson, better known for his mobster roles, looked appropriately intense as Paul Ehrlich in *Dr. Ehrlich's Magic Bullet.* In all of these movies, the experimental animals on whom discoveries relied definitely took second billing.

The synthesis of what became known as the sulfa drugs in the 1930s was the first real breakthrough in the struggle against bacterial infections. Before this occurred, some scientists had begun to believe that bacterial infections were somehow fundamentally different from those caused by spirochetes or protozoa and could not be chemically destroyed without also destroying the host— the infected human or animal. The German chemical giant I. G. Farben, a prominent maker of dyes, hired Gerhardt Domagk (1895–1964) in the late 1920s to direct research in bacteriology. In 1932, he found a dye that cured certain forms of streptococcal infection in mice. Domagk did not announce his dis-

Magic Bullets

Paul de Kruif's 1926 best-selling book *Microbe Hunters* spawned a number of other popular books and movies about medical research. In the photo, Edward G. Robinson as Paul Ehrlich confronts a chimpanzee in the 1940 Warner Brothers' production *Dr. Ehrlich's Magic Bullet.* The following passage conveys de Kruif's compelling style:

But Roux had got his start. With this silly experiment as an uncertain flashlight, he went tripping and stumbling through the thickets, he bent his sallow bearded face (sometimes it was like the face of some unearthly bird of prey) over a precise long series of tests. Then suddenly he was out in the open. Presently, it was not more than two months later, he hit on the reason his poison had been so weak before—he simply hadn't left his germ-filled bottles in the incubator for long enough; there hadn't been time enough for them really to get down to work to make their deadly stuff. So, instead of four days, he left the microbes stewing at body temperature in their soup for forty-two days, and when he ran that brew through the filter—presto! With bright eyes he watched unbelievably tiny amounts of it do dreadful things to his animals—he couldn't seem to cut down the dose to an amount small enough to keep it from doing sad damage to his guinea-pigs. Exultant he watched feeble drops of it do away with rabbits, murder sheep, lay large dogs low. He played with this fatal fluid; he dried it; he tried to get at the chemistry of it (but failed); he got out a very concentrated essence of it though, and weighed it, and made long calculations.

One ounce of that purified stuff was enough to kill six hundred thousand guinea pigs—or seventy-five thousand large dogs! And the bodies of those guinea-pigs who had got a six hundred thousandth of an ounce of this pure toxin—the tissues of those bodies looked like the sad tissues of a baby dead of diphtheria.

■ Quote from Paul de Kruif, *Microbe Hunters* (New York: Harcourt, Brace, 1926), 192. Illustration: From *Dr. Ehrlich's Magic Bullet* © 1940 Turner Entertainment Co. An AOL Time Warner Company. All rights reserved.

covery, which he called Prontosil, until 1935. A group of French scientists at the Pasteur Institute read Domagk's report and synthesized the drug themselves, then went further and found the active principle, which they called sulfanilamide. Both French and British scientists suspected that Domagk had isolated sulfanilamide by 1935, but that he did not announce it because it had already been synthesized in 1908 (although its antibacterial qualities had not then been recognized) and therefore could not be patented by his employer, I. G. Farben. Domagk's 1935 report was vague, and requests made by other scientists for samples of the drug to test were stonewalled for some time. By the late 1930s, however, clinical trials of Prontosil and sulfanilamide were so promising that other scientists sought to develop related compounds that might be of use in other bacterial infections such as pneumonia. The commercial success such a drug might enjoy was a motivating factor for several drug companies, and a group of drugs known as the sulfonamides, or sulfa drugs, began to appear.

Particularly before resistant strains of bacteria began to develop, sulfa drugs led to dramatic declines in the death rates for streptococcal infections, bacterial pneumonia, meningitis, and gonorrhea. But they proved to be of little use in local infections, such as those resulting from wounds or surgical operations. American university hospitals, with their close ties between clinical medicine and basic research, played a major role in the testing of sulfa drugs on humans; in this case, the German research institute, usually set up as a separate entity outside other institutions, was not as effective. Yet medical and popular enthusiasm for the sulfa drugs—labeled by the *New York Times* in 1938 as "the drug which has astounded the medical profession"—also led to some inappropriate and excessive use of the drugs by over-eager physicians and patients.* There were no established rules for clinical trials.

Sulfa drugs were hardly a cure-all, but their development demonstrated that chemical antibacterial drugs were possible. The London physician Alexander Fleming (1881–1955) had found in 1928 that the *penicillium* mold destroyed bacterial cultures on a plate of agar-agar gelatin. But he was unable to produce an active antibacterial principle from the mold, and applications of "mold-juice," which he named penicillin, to a few cases of human infection proved inconclusive. Although it did not harm mice or rabbits, he did not test its curative powers on infected mice. Ten years later, in Oxford, Howard Florey (1898–1968), Ernst Chain (1906–1979), and Norman Heatley (b. 1911) returned to the problem of the *penicillium* mold. By 1940, they had figured out how to produce enough of the purified drug to experiment on eight mice that had been

* *New York Times,* 20 November 1938, 5.

infected with streptococci. The four control mice died; the four mice injected with penicillin did not. Trials on seriously ill patients, although hampered by inadequate quantities of the drug, showed equally impressive results. A new antimicrobial drug had been discovered, one which, with its near relatives, ultimately changed the face of medicine.

By the late 1940s, the conquest of infectious disease seemed near, as more and more antibiotic drugs were developed. Most of these drugs, such as streptomycin and tetracycline, were, like penicillin, derived from natural substances including molds and soil bacteria. Among pharmaceutical companies, in the United States and abroad, competition was fierce to develop a new miracle drug that could be patented. As a result, thousands of bacteria and fungi were tested on thousands of animals to find just one new antibiological agent. In contrast to Ehrlich's 606 attempts, over seven thousand soil samples were tested in the late 1940s before a mold that produces the drug known as chlorampheticol was discovered. Competition, the push for profits, and public demand all fueled an enormous growth in the scale of research.

The antivivisection movement was still active in most European countries and in the United States, but during these years its influence diminished. Shaw's critique of vivisection in 1913 pointed out many of the weaknesses of the medical men's case, but by the 1920s, his opposition to vaccination and his noninterventionist views on medical care—views shared by many antivivisectionists—had begun to seem merely cranky in the light of the therapeutic successes of the period. Although the outcome of the bacteriological revolution was, as we have seen, mixed at best by the 1920s, the antivivisectionists' emphasis on physiological research and on the issue of pain began to seem beside the point.

Before World War I, antivivisectionists had gained much public support and had registered some notable public relations successes. Books such as Louise Lind-af-Hageby and Leise Schartau's *The Shambles of Science* (1903) revealed routine cruelty in medical classrooms and led to open conflict between medical students and antivivisectionists at the site of a statue of a brown dog in London. But these public relations successes were not translated into legislation, and Britain's law of 1876 remained the only attempt to regulate experimentation. Meanwhile, scientists had begun to recognize the need for public support. Criticism of the use of animals at the Rockefeller Institute, for example, led to a spirited response by its director of research. The American Medical Association founded a Council for the Defense of Medical Research in 1907, which challenged antivivisectionist claims. Its long-time chair, Walter Cannon (the chair of physiology at Harvard Medical School) formulated a code of conduct for animal laboratories that took much of the wind out of antivivisectionist sails. By the 1920s, a noted medical man claimed that antivivisection was "a lost cause."*

While antivivisection had lost some public support by the 1920s, it did not die out as a cause, and supporters continued to agitate for reform if not abolition of animal experimentation. In the United States, antivivisectionists focused largely on physiological experiments, and particularly on dogs. The use of pound animals and strays, and the fear that kidnapped pets were also used, were important elements of antivivisectionist campaigns from the 1930s to the 1960s. An article in *Life* magazine in 1966, which revealed horrendous conditions at so-called puppy farms—many of which sold animals to research laboratories—led to the first American legislation to regulate laboratory animal welfare.

Even when antivivisection was at its lowest ebb, historian Susan Lederer has shown that researchers were apprehensive about negative antivivisectionist publicity. Walter Cannon advised editors of medical journals to "eliminate expressions which are likely to be misunderstood" in articles they published, pointing particularly to animal experiments. He recommended that researchers should, for example, always mention when anesthetics were employed. The editor of the *Journal of Experimental Medicine* from the 1920s through the 1940s developed editorial guidelines emphasizing impersonal language ("animal" rather than "dog"), and often euphemistic medical terms that had the effect of distancing the reader from the act of experimentation. Illustrations might be of part of an animal rather than the whole. While this strategy may have minimized the grounds for antivivisectionist criticisms, it also fostered a culture of detachment from the experimental subject. As Lederer notes, "It is not without irony that the apparent linguistic insensitivity on the part of investigators may itself be a historical product of the bitter controversy with antivivisectionists."†

By the end of World War II, the successes of microbiology had gained it broad public support. Yet if science could be viewed as a savior, the explosion of the atomic bomb at Hiroshima in 1945 also revealed only too clearly its opposite role as a destroyer. The scale of scientific research grew immensely after the war. Did ethical questions therefore recede into the background? Some of the implications of these questions and concerns will be discussed in the next chapter.

---

*William Henry Welch, Dean of the School of Public Health at the Johns Hopkins University, in 1926: quoted in Susan E. Lederer, "Political Animals: The Shaping of Biomedical Research Literature in Twentieth-Century America," *Isis* 83 (1992): 62.

†Cannon quoted in Lederer, "Political Animals," 69; Lederer, "Political Animals," 61.

6 Polio and Primates

My kindergarten teacher in 1958 had had polio as a child and walked with the help of a heavy metal brace on one leg. I can still hear the clump of her step across the old wood floor of the classroom, and the click when she unlocked the brace to sit at the piano and sing with her class. These were not unusual sights or sounds in the 1950s. Everyone knew someone who had polio.

Polio seemed to be connected with summer and hot muggy days, its attack as sudden as thunderstorms and as ominous. One day you were out running around with your friends. Maybe you got a little overheated, and that night you got sick, with a fever and a sore neck. And maybe the next day you would be in an iron lung, struggling to breathe.

Polio, like smallpox, has disappeared so completely, at least in Western countries, that its absence seems normal. The last recorded case of naturally occurring polio in the Western hemisphere was in Peru in 1991. Along with hundreds of thousands of other children, I was injected with the Salk vaccine in the late 1950s, and a few years later lined up at school to get a dose of the Sabin oral vaccine out of a little paper cup. It tasted like sugar water. My young son tells me it now tastes like butterscotch.

The conquest of polio is one of the great triumphs of modern medical science. In the United States, the number of new cases of polio dropped from twenty thousand per year in the early 1950s (some fifty-eight thousand in the epidemic year of 1952) to five thousand in 1960, to virtually none a decade later. In 1988, the World Health Organization set as its goal the certification of the world as polio-free by 2005, and it is likely that this goal will be met. The Western Pacific, Europe, and the Americas have already been certified, and the number of polio cases worldwide fell by 95 percent between 1988 and 1999. Yet, as in most scientific accomplishments, the path to the polio vaccine was not straightforward or inevitable, and the conquest of polio was not without its costs. Competition among scientists to develop the vaccine—much like that in James Watson's account of the 1953 discovery of DNA, *The Double Helix* (1962)—showed scientists in a new and unfavorable light.

At least a million monkeys, and possibly up to five times that number, were sacrificed to develop the Salk and Sabin vaccines. In the light of new research about primate consciousness, intelligence, and sensibility, their use as experimental tools can no longer be taken for granted. The role of monkeys in the production of vaccines may also have led to the introduction of other viruses into the human population. The choice of experimental animal in the first decades of research on polio led to the construction of a seriously flawed model of the disease, which retarded its understanding for many years. In addition, the procedures followed in the early human trials of the Salk vaccine in the early 1950s raise questions about informed consent, human experimentation, and statistical design.

Poliomyelitis is a viral disease that, in its most severe form, attacks the lining of the spinal cord, causing paralysis. Although paralytic polio, even at its height, was an uncommon disease, its incidence among school-age children made it seem more common than it was. According to the severity of the attack, as well as measures taken during the attack, the paralysis can be temporary or permanent, can affect one or more limbs with varying severity, and can even paralyze the muscles of the chest and diaphragm, preventing breathing and swallowing. Polio can be deadly, but most patients have recovered, although some with permanent disabilities. Flaring to epidemic intensity around the world, its appearances in the twentieth century focused on the United States. In the 1952 epidemic, three thousand deaths occurred among fifty-eight thousand cases. The potential of lifetime disability, with its attendant social costs, made polio a much-feared disease. While it most often struck children—thus its alternative label of "infantile paralysis"—it could also affect adults. The New York politician Franklin Roosevelt was thirty-nine when he contracted polio in 1921, and his election as president of the United States eleven years later gave powerful impetus to polio research.

The poliovirus had existed for some time, but only in the 1890s did it emerge as an epidemic disease with the potential for causing paralysis. The irony of a seemingly new disease emerging at the height of the bacteriological revolution was not lost on scientists or the public. By 1910 scientists in Vienna and New York had established that a contagious virus caused polio, and that it could be passed from humans to monkeys. But no one at that time really knew what a virus was; unlike bacteria, viruses were invisible under conventional microscopes. The word *virus* simply signified an unknown infectious agent, not a specific entity. No one knew how infection occurred, or where the virus had suddenly come from. With polio, the new science of bacteriology had, temporarily at least, reached its limits.

Viruses were not visible to researchers until the development of the electron microscope in 1939. Unlike bacteria and other disease-causing microbes, viruses are not cells. Bacteria have their own DNA and independent metabolisms, and they can grow and reproduce (by division) on their own, as long as they are in a hospitable environment. Bacteria are susceptible to antibiotic drugs, which either kill them or keep them from reproducing. Antibiotics have no effect on viruses. Viruses are essentially a string of molecules that carry genetic information, surrounded by some sort of protective protein covering. A virus cannot reproduce on its own. It is a parasite that must have a host organism—a cell—in order to survive. Once it penetrates a cell, a virus can reproduce, sometimes very rapidly, using the host cell's proteins, by injecting its own DNA into the host's DNA. While many bacteria can infect a number of different species, the virus is usually much more species-specific, because it can be a parasite only on genetically compatible cells. The symptoms of viral infection can be caused by changes in the cells themselves, or by the body's reaction to the death of large numbers of infected cells. Over time, bacteria can mutate; that is, their genetic structure can change over time to the extent that, for example, they are resistant to certain antibiotics. Viruses can mutate too, and much more rapidly than can bacteria, since they have the ability to rearrange their genetic material immediately to suit their host.

We now know that the answers to the questions polio posed were even more ironic than early researchers suspected. The very public health measures that had helped to curb tuberculosis, cholera, and typhoid unleashed the scourge of polio. The poliovirus is carried in fecal matter, and in the days of open sewers and reeking outhouses, exposure to it at an early age was routine. Polio contracted in infancy often manifested itself merely in minor gastrointestinal symptoms. Many infants may have died of polio, but in an age of high infant mortality one disease was often not distinguished from another. Like many childhood diseases, it was endemic rather than epidemic, widely spread through the population rather than appearing in isolated flare-ups. Public health measures, based on the pre–germ theory connection between dirt and disease, led to closed sewers, clean water, and cleaner food and ended this routine exposure to and infection by the poliovirus. Children, rich and poor, born after the turn of the century never developed those antibodies that could help them resist a more severe attack of the virus. Franklin Roosevelt, born in 1882, was an exception: his privileged childhood had limited his exposure to the sources of disease. As a result, in the twentieth century polio struck the unprotected population with unprecedented virulence.

The polio epidemic of 1916 was a shock to the nation: over a period of five months, twenty-seven thousand people, mostly children, contracted the disease, and six thousand died. New York City was an epicenter of disease, with almost nine thousand cases and twenty-four hundred deaths. Theories of public health based on cleanliness were put to the test, for unlike cholera and other epidemic diseases, this disease did not discriminate between clean and dirty neighborhoods, between rich and poor, urban and rural. Although the germ theory, by establishing impure food and water as carriers of disease microbes, had helped to explain the efficacy of cleanliness in preventing disease, it could not explain this new plague.

With confidence born of the successes in bacteriology of Pasteur and Koch, a few scientists turned to research on polio. But other than acknowledging that polio was caused by a virus, not a bacteria, scientists in the early twentieth century agreed on little else. Even diagnosis was difficult, since in its earlier stages, polio resembled many other diseases. The spinal tap—obtaining a sample of spinal fluid by means of a puncture—began to be widely used in 1916, but it was a difficult procedure and did not lead to a foolproof diagnosis. Few hospitals and fewer doctors had the expertise or equipment to analyze spinal fluid for the presence of a virus undetectable under a microscope.

Initially, polio research followed what had by then become the standard procedures in bacteriological research. Researchers attempted to isolate the disease organism and, following Koch's postulates, to experimentally produce the disease in animals. But there was an important difference between polio research and research on other diseases: unlike most other research at the time, polio research relied heavily upon the use of monkeys, because only humans and monkeys, it seemed, could get the disease.

Despite the fact that Charles Darwin had established the close evolutionary proximity between humans and other primates, and Galen had long ago recognized their anatomical similarities, primates were used infrequently in research before the late nineteenth century. Claude Bernard once used a monkey in an experiment, but he refused to do so again because of its resemblance to humans. Later, in the 1870s, the British physician David Ferrier (1843–1928) undertook studies on the localization of brain functions in monkeys, and antivivisectionists criticized him heavily. Unlike dogs or the favored rats and mice of today's laboratories, monkeys were difficult and expensive to obtain. In 1908, however, Karl Landsteiner (1868–1943) and Erwin Popper in Vienna accomplished an experimental breakthrough. By injecting material from the spinal cord of a polio victim into the spines of two monkeys, they induced the symptoms of paralytic polio in the animals. Subsequent autopsies revealed polio-like changes in the monkeys' spines. Landsteiner and Poppert established two important principles that guided future polio research: that monkeys could be used as models for human illness, and that the seat of polio was in the spinal cord. Neither of these principles, it turned out, was entirely flawless.

Chief among American polio researchers in this period was Simon Flexner (1863–1946), laboratory director of the Rockefeller Institute for Medical Research. Flexner believed that the laboratory was where a cure for polio would be found, and in response to antivivisectionists in 1911 he announced that a cure was imminent. Flexner's high status disinclined other researchers to enter the field or, later, to challenge his conclusions; and his focus on the laboratory, like Claude Bernard's, excluded the evidence of practicing physicians. Like Landsteiner, Flexner chose the rhesus macaque as an experimental animal. Because monkeys were not yet bred for research, all of them were imported from south Asia. Both his choice of subject and his method of inducing the disease into the animal by means of injection into the brain or spine had unforeseen consequences.

Because Flexner and his research associate, Hideyo Noguchi (1876–1928), could not see the poliovirus, it could be detected in experimental animals only if they exhibited symptoms. The only way to induce symptoms in a rhesus monkey was through the brain, and the symptoms produced centered on the brain and spine. This led Flexner to conclude that polio was a disease of the central nervous system. But if he had used other primates, his conclusion might have been quite different, for other species, including humans, can develop polio by ingesting the virus. We now know that symptoms of oral ingestion—the way most people contracted polio—centered first on the digestive system, and only later, after the virus reached the blood, perhaps on the spinal cord. Paralytic polio, in other words, was the exception, not the normal course of the disease. Flexner's insistence on the involvement of the nervous system led him to discount reports of physicians who mentioned stomach upset as a symptom, and led him to view paralytic polio as the norm.

Flexner experimented on polio for several years, dominating the field; and like Pasteur, he passed his viral material through a number of animals. Because viruses are highly adaptable microorganisms, Flexner's poliovirus adapted itself perfectly to its host, the rhesus monkey. Unfortunately, this also made his research distinctly less applicable to humans. In Flexner's macaques, polio became a virus that affected only the central nervous system. As far as Flexner and Noguchi could determine, the laboratory animals displayed none of the gastrointestinal symptoms clinicians had sometimes noted in human cases. The adaptation of Flexner's virus, therefore, supported his assumptions.

Owing to Flexner's prominence as a researcher, his conclusions were taken as fact by physicians and by the public. Flexner's papers include many letters from distraught parents begging for a cure. His insistence that the poliovirus was introduced in humans only via the nose and mouth, from whence it traveled directly to the brain, led to such remedies as useless nasal sprays. His emphasis on paralysis and other neurological symptoms led physicians to ignore nonparalytic cases, which greatly confused the picture of polio's spread and distribution, leaving the means of contagion unexplained. Because Flexner believed that the blood did little to send the virus throughout the body, for many years physicians and researchers disregarded the blood and its antibodies. One historian comments that Flexner's immense influence in the United States "may have delayed understanding of polio's complex clinical and epidemiological features for at least a generation."*

The Race for a Vaccine

Franklin Roosevelt's polio helped to change the image of polio from a disease of children to one that could strike anyone at any time. Roosevelt's political prominence, personal charisma, and numerous connections made polio research a national priority. The National Foundation for Infantile Paralysis, founded with his sponsorship in 1937, established the "March of Dimes" fundraising campaigns, which raised millions of dollars for polio treatment and research.

Researchers from the 1920s onward began to dismantle Flexner's interpretation of the disease. While rhesus macaques remained the preferred experimental animal, in part because of their availability (they were readily imported from India), other species of monkeys and chimpanzees, whose experience of polio more closely mirrored the human, were also used. Laboratory researchers began listening to physicians and epidemiologists, who then made connections between exposure to polio (even if it led to no symptoms, or mild ones), antibodies in the blood, and immunity, although there continued to be little understanding of how the immune system worked. Different strains of the poliovirus were studied. Each laboratory began with human polio material that they kept growing in a series of monkeys. But someone soon noticed that a monkey infected with one lab's batch of polio was not necessarily immune from infection from another lab's batch. Obviously, different strains (what we would call genetic variants) existed, although no one knew how many. In the late 1940s, a program to establish the number of strains, or to "type" them, established that there were three.

* Naomi Rogers, *Dirt and Disease: Polio before FDR* (New Brunswick, N.J.: Rutgers University Press, 1992), 77.

Meanwhile researchers used many thousands of monkeys. Until the 1930s, virologists could not grow viruses outside a living medium. So monkeys became test tubes, growing the virus in their spines. The monkeys were killed, their spines were ground up, and the process was repeated. The population of rhesus macaques in their native India plummeted from an estimated five million to ten million in the 1930s to fewer than two hundred thousand by the late 1970s. Polio research, especially after World War II, contributed to this decimation, and it is worth noting that the United States imported two hundred thousand rhesus macaques a year in the 1950s for vaccine production and research. India began to restrict their exportation in 1955. Already by the 1930s it was difficult to import enough healthy animals to keep research programs going, and the National Foundation for Infantile Paralysis developed an animal facility at Okatie Farms in South Carolina, which provided animals, including monkeys, for projects funded by the foundation.

The development of tissue culture techniques in the 1930s lessened researchers' reliance on animals, although it did not eliminate it. The ability to grow a virus in a sterile medium in a laboratory flask, rather than in an animal's body, opened the possibility for wide-scale production of a polio vaccine, which would otherwise require additional millions of monkeys. Early antibacterial drugs such as sulfonamides lessened the possibility of bacterial contamination of the medium. In 1936, Albert Sabin (1906–1996) and Peter Olitsky (1886–1964) grew the poliovirus in a test tube, using human embryonic nervous tissue as their medium. Their use of nervous tissue indicated that they, following Flexner, believed polio to be a disease of the nervous system. Because attempts to grow the virus in other kinds of tissue failed, many scientists continued to assume that Flexner was right. But as Pasteur's rabies vaccine had demonstrated, a vaccine developed from nerve cells carried the risk of fatal brain diseases and allergic reactions. In addition, while embryonic human tissue was the ideal medium for tissue cultures, it was not easy to come by, and the use of such material, obtained from stillbirths and miscarriages, remains controversial. Monkeys were the inevitable substitute, and work on monkeys continued.

Public pressure to develop a polio vaccine was enormous, and whoever developed the vaccine would undoubtedly become famous. Two researchers in the mid-1930s announced they had developed a vaccine. Maurice Brodie (1903–1939) of New York University employed virus from monkey spinal cords which he killed with formalin, a solution of formaldehyde. Killed-bacteria vaccines had been used for some time (such as the typhoid vaccine, developed in the 1890s), but no one was certain that a killed virus could induce immunity. Brodie and William H. Park (1863–1939), director of the laboratories of the New York Board of Health, tested the vaccine on twenty monkeys, as well as on Brodie and some of his laboratory assistants, none of whom experienced ill effects. Brodie proposed a field test—that is, a test outside the laboratory—on three thousand children. His rush to proceed with human trials was prompted by his rivalry with Philadelphia researcher John Kolmer (1886–1962), who was developing a live (but weakened) virus vaccine. Brodie went ahead to vaccinate sixteen hundred California children and later, several thousand others. He was forced to stop when several children developed allergic reactions. It soon became clear, moreover, that the vaccine did not confer immunity. Kolmer also tested his vaccine on himself and on his two children before vaccinating some ten thousand children. But several of these children developed severe cases of polio and some died, and Kolmer withdrew his vaccine.

Brodie and Kolmer stimulated an ongoing debate about killed- versus live-virus vaccines. Most virologists in the 1930s believed that a successful vaccine had to follow the Jenner model: a weakened version of the disease. The poliovirus was generally attenuated by passes through animals, or later, through tissue cultures, although Kolmer used chemicals to weaken the virus. In all live-virus vaccines, there is a danger (not fully recognized at the time) that the weakened virus will back-mutate into a more virulent form. This could not happen with a killed virus, but scientists believed that a killed virus could not generate lasting immunity. The nature of immunity was still unclear. Peter Olitsky and Herald Cox (1907–1986) tested the Park-Brodie vaccine in monkeys at the Rockefeller Institute, and announced that it produced antibodies in the blood but that this did not constitute true immunity. They believed antibodies were products of infection, not evidence of immunity.

The failures of Brodie and Kolmer were much in mind two decades later, when Salk's killed-virus vaccine began to be tested. But some of the issues of consent that their tests raised remained unresolved. Brodie and Kolmer insisted that they did not vaccinate anyone without their (or their parents') consent. But what did the parents consent to? How much did they understand of the experimental process? Antivivisectionists pointed to the use of orphans in the trials as a blatant example of the lack of consent. Park argued that because orphans were not used until the vaccine had already been tested on other volunteers, it was no longer experimental. When, indeed, did field trials stop being experimental and start being therapeutic? How many cases added up to proof?

The quest for a polio vaccine intensified in the years after World War II. The National Foundation for Infantile Paralysis had, from the beginning, used

Hollywood figures and sophisticated fundraising techniques. In postwar America, the March of Dimes poster children served as constant reminders to the public of the importance of conquering polio. A significant research breakthrough had occurred in 1939, when a scientist with the U.S. Public Health Service found a strain of the poliovirus known as the Lansing strain, which would grow in the brains of mice and rats. In 1948, John F. Enders (1897–1985), Thomas H. Weller (b. 1915), and Frederick C. Robbins (b. 1916) in Boston grew the Lansing strain on non-nervous tissue, opening the door for the production of a safe vaccine. Enders, Robbins, and Weller won the Nobel Prize for Medicine in 1954 for this work.

Although tissue culture lessened the reliance on animals, it did not end it. In fact, large-scale vaccine production in the 1950s increased the number of monkeys used. Enders had used human tissue for his cultures: discarded foreskins from circumcised baby boys. But as we have seen, the use of human tissue was controversial and its supply was limited. Tissue-culture techniques meant, however, that material from one monkey could now produce two hundred culture tubes. Previously, a single monkey spine had yielded only three samples of virus.

In 1947, the National Foundation for Infantile Paralysis funded a young researcher, Jonas Salk (1914–1995), of the University of Pittsburgh, to work, along with researchers at several other laboratories, on identifying the different strains of the poliovirus. At New York University and the University of Michigan, Salk had worked with his mentor, Thomas Francis (1900–1969), who had developed a killed-virus vaccine for influenza. The work of Salk and others confirmed that there were only three strains of the poliovirus, but this knowledge was hard won in a lengthy process requiring thousands of monkeys. The procedure involved infecting a monkey with a known strain, waiting for it to recover, then introducing an unknown strain. If the monkey got ill again, the unknown strain was different from the known one. But that did not in itself identify the unknown strain, which then was further challenged by another unknown. Complicating the procedure were variations in virulence of the viruses: how much of the virus did the scientist need to give to the monkey to make it sick, without making it die? Monkeys were expensive, and researchers did not like to lose monkeys unnecessarily.

Salk was very good at caring for monkeys, which were delicate organisms, easily injured and quick to catch whatever disease was in the air. In his laboratory in Pittsburgh, he designed housing for the experimental monkeys. Salk even donated some surplus animals to a local zoo. But it would probably be a mistake to see Salk as being exceptionally concerned about his animals. Because monkeys were expensive and essential to the experimental process, they had to be treated well, at least from the point of view of physical well being.

In order to save monkeys, which meant to save money as well as time, Salk turned to tissue cultures in his virus-typing research. Typing a virus in a test tube took only a few days rather than weeks. The tissue-culture technique could also be used to develop a vaccine, which Salk now set out to do. He chose a killed-virus vaccine, in which the virus was killed chemically but with its antibody-producing qualities intact, versus a live-virus vaccine in which the virus was weakened but still alive. Brodie's failure with a killed-virus vaccine did not deter Salk, and it turned out that Brodie had used too much formalin in his own preparation, so that the killed virus did not retain enough virulence to promote antibody formation. But there was also a danger of leaving live, toxic virus in such a vaccine. The only way to determine whether a killed virus was truly inactivated was to inject it into monkeys' brains. But animals varied in their susceptibility to polio, so a virus might have no effect on one species and yet be toxic to another. Thomas Francis was not certain that his killed-virus influenza vaccine conferred long-lasting immunity. Many continued to doubt that a killed-virus vaccine, which would be safer because it could not cause illness, would actually work. Although the relationship between antibodies and immunity was clearer by 1949 than it had been in the 1930s, even in 1953 Salk could refer to the relationship between antibodies and disease resistance as an "assumption," not a proven fact.* Other scientists, including Sabin, Cox, and Hilary Koprowski (b. 1916), later developed live-virus vaccines, but in the early 1950s, only Salk and Isabel Merrick Morgan (1911–1996; reported successful monkey trials in 1951) worked on a killed-virus vaccine.

As early as 1950, Salk was suggesting human trials for his vaccine, based on his success with vaccinating monkeys. He proposed to the head of research for the National Foundation that experiments might be tried on chimpanzees and also on mentally retarded and institutionalized children, as well as on prison volunteers. But the foundation was unconvinced and suggested he finish the virus-typing they had funded him to do. Koprowski, meanwhile, who worked for a commercial drug company, tested his live-virus vaccine on twenty retarded children at an institution in New York State. Two years later, Salk proceeded with human trials. Several laboratory workers, and Salk himself and his children, had been vaccinated without incident. Salk next tested polio victims

---

*Jonas Salk, "Principles of Immunization as Applied to Poliomyelitis and Influenza," *American Journal of Public Health* 43 (November 1953): 1384–98.

at a home for handicapped children. If the vaccine raised the level of blood antibodies in the children, Salk could claim that it was more effective than natural polio in inducing immunity.

He then turned to a Pennsylvania school for the mentally retarded as the first site for his vaccine trials. The mentally incompetent had long been used as experimental subjects. In the 1660s, early blood-transfusion experiments had been performed on such individuals. More recently, among those chosen by Hideyo Noguchi in 1911 to test his new diagnostic serum for syphilis were residents of a New York asylum.

The residents of the Polk State School were not competent to give consent to Salk's testing. In this period, consent was an affair between the experimenter and his subject: there were no institutional review boards, no protocols to be reviewed. After 1946, the American Medical Association required that volunteers consent to experimentation, but this requirement had no legal standing. In 1952, lawyers for the National Foundation and the State of Pennsylvania, who ran Polk State School, negotiated the terms of the consent forms, which were signed by the legal guardians of the children (in many cases, this was the state), and Salk began tests in May 1952. In the course of these small-scale trials, Salk continued to refine and modify his vaccine.

Few knew of Salk's work in 1952. He announced the results of these trials at a meeting of polio researchers in January 1953, and several participants were impressed and suggested a much larger field trial. Albert Sabin, supported by John Enders, objected, saying Salk needed at least ten more years of research to prove that the killed-virus vaccine was effective and to determine the proper dosage. To Sabin, who was pursuing the more mainstream science of a live-virus vaccine, the killed-virus, tissue-culture vaccine of Salk was just too risky. The National Foundation, eager for a vaccine, promoted the idea of a field trial, and enlisted public opinion on its side.

The 1954 Field Trials: Human Experimentation?

Salk published the results of his 1952 trials in the *Journal of the American Medical Association* in March 1953. But even before publication, the National Foundation spread news of the new vaccine to the press, including an article in *Time* magazine featuring Salk. The National Foundation promoted his work and sought support for large-scale field trials. Salk was interviewed on national radio, explaining his research and attempting to quell rumors that a vaccine would be immediately available. A March 1954 article in *Time* magazine focused on Salk, who was characterized as "a young man in a hurry." This publicity made Salk highly suspect to the rest of the scientific community, to whom publication and evaluation by peers was the only way to assess research. In the eyes of his fellow scientists, bringing the public into it was a mistake that challenged scientists' autonomy and endangered scientific advance.

Salk steadily continued small-scale field trials in 1953 while also expanding his laboratory. Okatie Farms sent Salk fifty squirrel monkeys a week, and his lab also used thousands of mice, as well as chickens. Although tissue cultures meant that Salk used far fewer monkeys than many of his predecessors in polio research, his discovery that monkey kidney tissue produced the best cultures meant that thousands of monkeys continued to be used. Production of enough vaccine for a large field trial, not to mention enough to inoculate every child in the United States, would take many thousands more.

The National Foundation appointed Thomas Francis to oversee the national field trials, scheduled to take place in 1954. Francis favored a double-blind trial, in which half of the children would be injected with a placebo. The subjects would be coded so that no one would know who got the placebo and who the vaccine until all the results were tabulated. In terms of experimental models and statistical rigor, a double-blind test was by far the best. But to be successful such a trial must also be randomized: the choice of who received the new vaccine and who received placebo would be entirely random, although children in the same classes and of the same age had to be matched against each other, since geographical variation in the natural incidence of polio was high. The statistical task was daunting, especially since randomized vaccine trials had never been done before. The first randomized and double-blind trial of any sort had been performed in the United Kingdom only five years previously.

Side by side with the double-blind test, which took place in eleven states, was an "observed control" test in thirty-three states in which the control group was not given a placebo, but merely observed. While the randomized placebo model was essential to eliminate the effects of bias on the results and would help to placate the doubts of Salk's critics, in his words it "would make the humanitarian shudder."* But nonparalytic polio looked a lot like a number of other flu-like diseases, and a doctor could well make a wrong diagnosis if he knew his patient had been vaccinated. The observed control trials, however, were far easier to sell to nervous parents, who disliked not knowing whether their child had received the vaccine. The fear of polio was so great that parents overwhelmingly agreed that their children should be "polio pioneers" and participate in the trials. The consent forms they filled out did not employ the word *consent*. Over 60 percent of parents requested that their children partic-

* Marcia Meldrum, "'A Calculated Risk': The Salk Polio Vaccine Field Trials of 1954," *British Medical Journal* 317 (1998): 1233–36, at 1234.

Polio Field Trials

In explaining the decision parents made to allow their children to take part in the wide-scale field trials of the Salk vaccine in 1954, historian Jane Smith has vividly portrayed the near-panic of parents at the time of the 1953 meeting of the National Foundation for Infantile Paralysis. The polio epidemic of 1952 had attacked in every state; the daily toll of polio cases was even reported on television, as deaths in the Vietnam War would be a decade later. Smith links the fear of polio to the pervasive fear of nuclear holocaust. In what she calls "an atmosphere of mass vulnerability," parents welcomed early news of the Salk vaccine. They trusted science and technology to solve modern problems, and they trusted in progress. In Smith's words, "People who had accepted the plausibility of Spam

were not likely to doubt the safety of a polio vaccine sponsored by the National Foundation for Infantile Paralysis."

■ Quotes from Jane Smith, *Patenting the Sun: Polio and the Salk Vaccine* (New York: Morrow, 1990), 160, 162. Illustration: Jonas Salk holding flasks of vaccine, ca. 1954. Courtesy of the Jonas Salk Trust.

ipate in the placebo trials, and almost 70 percent in the observed control trials. Extensive publicity and in-school presentations encouraged parents to understand the benefits of what was about to occur.

Coordinating a field trial of 1.6 million children was an enormous task. It included monitoring several commercial drug manufacturers to ensure vaccine production, which claimed the lives of four thousand monkeys per month in 1954, in part because each batch of vaccine was tested three times. The field trials began in April 1954. The "Polio Pioneers," more than six hundred thousand first, second, and third graders in several states, were injected with either the Salk vaccine or a colored-water placebo, and returned for two more shots. Others had blood samples taken, and over a million children participated as observed controls. At the end of the trials in June, records of the inoculations were sent to Francis in Michigan for evaluation. In December, Albert Sabin began his own trials, on twelve prison volunteers, of his live-virus vaccine.

On 12 April 1955, Francis announced that the trials of the Salk vaccine had been successful, displaying an effectiveness of 80 to 90 percent in preventing paralytic polio. On the same day, the secretary of health, education, and welfare licensed the vaccine for distribution. However, the move to mass production of the vaccine was not without its hurdles. Making the vaccine was a complicated process, and the drug companies assigned to make it had to exercise special care. Only a few weeks after Salk's vaccine was licensed, some children who had been vaccinated came down with polio. The illness was traced to a batch of improperly prepared vaccine that contained live virus.

Meanwhile Sabin and Koprowski continued their work on live-virus vaccines. In the late 1950s, each conducted field trials—Koprowski in what was then the Belgian Congo, and Sabin in Russia. In 1961, the Sabin live-virus vaccine was licensed for use in the United States, and by 1965 it had virtually supplanted the Salk vaccine. The Sabin vaccine, taken orally, confers protection by means of the intestines rather than the blood and is therefore more effective in areas of epidemics. In addition, contact with a Sabin-inoculated person could provoke immunity by means of infection even in someone who had not been vaccinated. The Sabin vaccine had the additional advantages of being cheaper to produce and easier to administer. However, it always held the slight danger that the live virus it contained could mutate into a more virulent form.

The Monkey Connection

In June 1999, a federal advisory panel recommended that the Sabin oral polio vaccine be abandoned in favor of the Salk vaccine. What caused this dramatic change in policy? It has long been known that live-virus vaccines carry the risk of transmitting the disease they are meant to prevent. Even if the virus does not back-mutate into a more virulent form, the weakened version can in some cases be sufficient to cause disease. Because the Sabin vaccine contains live virus, it therefore may induce the disease, and a few children every year in the United States have contracted polio from the Sabin vaccine. Although the risk is low, one case for every 2.4 million doses, it is real. Because natural polio no longer exists in the United States, elimination of vaccine-caused polio now seems a reasonable goal.

Other risks associated with the polio vaccine have nothing to do with the poliovirus, but with the use of monkeys to make it. Not only do human viruses such as polio affect monkeys, but monkey viruses can also affect humans. In the 1930s, a researcher was bitten by a monkey and died from a mysterious disease. Albert Sabin and Arthur W. Wright (1894–1976) proved the death was caused by a monkey virus, which they named B virus, the initial of their dead colleague. B virus was isolated in the 1940s, and fortunately for Salk, the formalin he used to kill the poliovirus in his vaccine also killed B virus. But other,

The Polio Research–AIDS Connection

Most scientists agree that AIDS had its origin in a monkey virus that at some point was introduced to the human population. Could polio research and the polio vaccine have been responsible? A 1992 article in *Rolling Stone* magazine suggested just this connection. *Rolling Stone* is better known for its rock-and-roll coverage than for its medical criticism, but the article led to much discussion in the scientific community as well as among the public. The writer, Tom Curtis, linked the tests of Hilary Koprowski's live-virus oral polio vaccine in the Belgian Congo with the outbreak of AIDS in that area. Unlike Jonas Salk, who mainly used rhesus macaques to develop his vaccine, Koprowski used, among other species, African green monkeys, which were easier to obtain after India began restricting exports of rhesus macaques in 1955. African green monkeys are known carriers of SIV, the simian immunodeficiency virus, which most epidemiologists recognize as the monkey counterpart of HIV, the virus that causes AIDS.

Koprowski's oral vaccine was sprayed into the mouths of some 325,000 people between 1957 and 1960. According to later epidemiological studies, the first outbreaks of AIDS took place in the Congo in 1959. In 1999, researchers announced they had found the origin of the AIDS virus not in green monkeys but in a particular species of chimpanzee. This seemed to rule out the polio connection, but Edward Hooper's

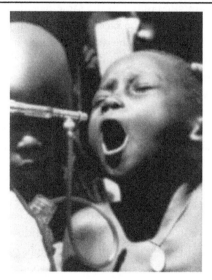

book *The River: A Journey to the Source of HIV and AIDS,* which appeared at the end of 1999, raised the issue all over again. Hooper suggested that chimpanzee tissue might have been used to prepare a batch of the vaccine. Although his case was circumstantial, it revealed, among other things, that no records were kept of what primate species were used. Koprowski has denied that he ever used chimpanzees, but Hooper's book raised enough questions that investigation is ongoing, including testing of samples of Koprowski's original vaccine. In a further complication, researchers reported in January 2002 that no wild chimpanzees were found to be infected with AIDS-like viruses.

■ Child receiving Koprowski's oral polio vaccine, Belgian Congo, 1957. Photograph by Robert Phillips.

unknown viruses remained in the monkey tissue used to make the vaccine. At the 1953 meeting in which Salk had announced his vaccine, some scientists worried about his use of monkey kidney tissue in terms of immune reactions, but not as a carrier of other viruses. In the early 1960s, a virus known as SV-40 (for simian virus number 40) was discovered in the Salk vaccine. It had survived the formalin and was injected into millions of Americans between 1954 and 1961. No one apparently became ill from SV-40, although recent studies have linked the virus to certain cancers. But the incident raises troubling questions about our use of animals, especially primates, to produce vaccines.

Primates' bodies have been used not only to produce vaccines but also in transplant surgery. In 1992, surgeons transplanted a baboon liver into a human. The famous case of Baby Fae, in which a baboon heart was transplanted into a human baby, took place in 1984. The medical, not to mention the ethical, issues of disease transmission from primates to humans have been largely ignored, despite the heavy publicity given to new tropical fevers such as Ebola and Marburg, which apparently have traveled from monkeys to humans.

The Right Model

Because primates, especially apes, are similar to humans in many ways—chimpanzees share 98 percent of the human genetic code—they have been viewed, especially in the twentieth century, as ideal experimental models for human medicine. Polio research is just one example of the use of monkeys and apes in disease research. Their similarity to humans has also meant that primates have been the preferred experimental model in psychological testing. Little protest was raised against the use of monkeys in polio research. The case of psychology is vastly different. By emphasizing the mental and emotional similarities between humans and primates, psychological research has seemed much more ethically problematic to nonscientific observers.

One of the best-known examples of the use of primates in psychological testing was the work of University of Wisconsin researcher Harry Harlow (1905–1981) between the 1930s and the 1970s. Harlow's studies of the mother-infant bond made him one of the most influential, and most controversial, figures in modern science. His research was based on the premise that animals, and particularly primates, have higher levels of cognition than most researchers of the time believed. In the 1930s, much animal work, including the classic studies of rats in a maze, was based on the view that animals could learn only in small increments, in terms of a single stimulus and a response. Harlow, in contrast, argued that primates were capable of complex problem solving, forming concepts, and using tools. The driving mechanism of learning was not simply the satisfaction of earning the reward, the food at the end of the maze. Curiosity and solving the problem were themselves motivations. Harlow's continuing study of brain function suggests that he sought a physiological basis for behavior, but he never fully worked this out. In contrast to prevailing behaviorist theories, which asserted the dominant influence of the environment on behavior, Harlow argued that some behaviors were innate rather than learned.

With such a favorable view of primate intelligence, Harlow should, it seems, have been a modern hero rather than a villain. During the period when Harlow conducted his research on maternal deprivation, monkeys were subjected in the name of science to radiation exposure, space flight, blunt impact trauma, vertigo, gunshot wounds, and a number of diseases. But Harlow's research, which explored one of the most basic of relationships, eventually made him a touchstone for the new animal rights movement. Like Magendie in the nineteenth century, Harlow became a model of the evil scientist.

Harlow began his work with rhesus macaques in the 1930s. Eventually, he set up one of the first breeding colonies of laboratory monkeys. In some of his intelligence experiments, he would take infant monkeys from their mothers at birth, in order to chart their learning process. Although the babies thrived physically, they showed strong behavioral anomalies, and by the 1950s Harlow had turned to the mother-infant bond as the subject of his research. He began to build surrogate mothers: first soft, cuddly ones, but later cold, hard, vicious ones. He found that babies preferred soft "mothers" to hard wire ones. Even when the wire "mothers" provided all the food, the baby macaques preferred comfort. But given the choice between a cold, wire "mother" or no mother at all, the babies chose any mother, even one that blasted it with cold air or stuck it with spikes. Further experimental variations showed that neonatal monkeys preferred warm wire mothers to ones that were soft but cold.

Plainly, infant macaques craved contact with a mother. What would happen if they were brought up without it? Infant macaques brought up in isolation were fearful and nonsocial, in contrast to the naturally social behavior of macaques. They did not know how to function sexually, and if females became mothers (which Harlow accomplished by tying the females down so that the males could impregnate them), they were terrible mothers. The mothers who had been raised in isolation had no idea how to nurture. They were unresponsive to indications of discomfort, and even hurt their infants.

Harlow pushed his research further and further, with multiple variations on the theme of isolation—from their mothers, from each other, and eventually isolation from any vestige of a supporting environment—producing mon-

The Rearing of Infants

Harry Harlow concluded his presidential address to the American Psychological Association in 1958, which described his maternal-deprivation experiments, as follows:

The socioeconomic demands of the present and the threatened socioeconomic demands of the future have led the American woman to displace, or threaten to displace, the American man in science and industry. If this process continues, the problem of proper child-rearing practices faces us with startling clarity. It is cheering in view of this trend to realize that the American male is physically endowed with all the really essential equipment to compete with the American female on equal terms in one essential activity: the rearing of infants. . . . But whatever course history may take, it is comforting to know that we are now in contact with the nature of love.

■ Quote from Harry F. Harlow, "The Nature of Love," *American Psychologist* 1 (1958): 685. Illustration: An infant monkey avoids a soft but ice-cold mother surrogate. In H. F. Harlow, M. K. Harlow, S. J. Suomi, "From Thought to Therapy: Lessons from a Primate Laboratory," *American Scientist* 59 (1971): 541. Courtesy

keys with more and more abnormal behaviors. In this context, the ultimate experimental situation was what Harlow called "the pit of despair." This was a wedge-shaped metal chamber, deep enough so that the monkey saw only the hands of a caretaker, slippery enough so that climbing out and basic movement was all but impossible. The pits of despair created monkeys incapable of acting like monkeys, monkeys that were permanently psychotic, so demoralized that they simply huddled, day after day, in the bottom of the pit. Harlow suffered from depression himself, and the image of being sunk in a well of depression may have influenced this experimental model, although in this case Harlow seems to have confused being depressed with the forces that cause depression.

Harlow made genuine contributions to psychological research and to child psychology. Early-twentieth-century manuals of child rearing warned against excessive contact between mother and child, for fear of "spoiling" the child. Even Dr. Benjamin Spock, whose popular handbook on baby and child care

Model Mice

American scientists use about forty million mice per year, and mice and rats together represent 80 to 90 percent of all laboratory animals. In many ways, they are the ideal experimental animals. They are easily tamed, and their small size makes housing and feeding cheap and simple. Their prolific reproduction habits make it easy to develop a mouse colony, and because they breed several times a year, they are especially suitable for genetic experiments that require tracing several generations. In addition, mice get many of the same diseases humans do. They can be infected with syphilis or salmonella, or have tumors or electrodes or catheters implanted in them.

Robert Koch and Paul Ehrlich used mice and other small mammals in disease research at the end of the nineteenth century. But the mice used today are highly specialized research tools, far different from Ehrlich's. They are specially bred for particular characteristics such as obesity. Techniques of genetic engineering have created even more specialized mice, such as "knockout" mice in which a particular gene is "knocked out" and transgenic mice whose genomes contain human DNA. The use of inbred mice (bred within a closed community to emphasize certain genetic factors) began around 1914. Inbreeding assured standardization—that all the mice were essentially alike. Certain inbred mice displayed greater susceptibility to cancers, and by the 1930s, mice were central to cancer research.

The cancer researcher Clarence C. Little opened the Jackson Laboratory in Maine in 1929. This laboratory was so successful at developing inbred mice that it began to market them in the mid-1930s. Today, researchers can leaf through catalogs of mice and other research animals and order exactly the right animal tool for research. The Jackson Laboratory today maintains 2,500 varieties of mice, and in 1999 it shipped 1.7 million mice to labs around the world. Unlike dogs and cats, mice have seldom been the focus of humanitarian concern, and the American Animal Welfare Act does not include them among regulated animals. Many seem still to agree with Little, who found the experimental use of mice a means to make a pest socially useful.

■ Laboratory mouse, "Lucky." Courtesy of Dennis O. Clegg, University of California, Santa Barbara.

first appeared in 1946, warned against coddling in early editions of his book. Harlow was in part inspired by John Bowlby's research on maternal depriva- tion among relocated British children during World War II. After Harlow's research, no one could doubt the critical importance of close mother-infant contact in the early months of life. Subsequent research has detailed physio- logical changes, including the production of particular growth hormones, in baby monkeys who have maternal contact versus babies who do not.

But animal activists, and even some researchers, believe that Harlow at some point crossed a line between curiosity and cruelty, and ultimately be- tween good science and bad science. But where was the line? Harlow's flippant language in his reports often makes the reader cringe: not only were his isola- tion chambers "pits of despair," but he also referred to the "rape rack" he used to immobilize isolation-reared females for insemination, and to "hot mamas." He displayed no concern in the face of the most cruel experiments, although the impact of the experiments was precisely in their emotional power. Was it his penchant for publicity? Which experiments crossed the line? A laboratory inspector criticized Harlow's "pits of despair" as inhumane, but his work con- tinued to be funded and published in respected journals, and colleagues and students at the time—whatever they may have felt later—did not protest. His prominence as a researcher, which included election to the National Academy of Sciences and the American Philosophical Society as well as the presidency of the American Psychological Association, shielded him from criticism. And what of his other isolation experiments? Why were the pits of despair worse?

Perhaps the cumulative effects of Harlow's experiments are what horrify many who look back on them today. One of his former students, John Gluck, has analyzed some of the reasons that Harlow's work was not criticized by those closest to it in its day, and this analysis is relevant to many other laboratory sit- uations. For his students, Harlow modeled research as a war with nature, a con- stant struggle, and, as in war, normal rules of morality are suspended. There are many examples, from Galen to the present, of military metaphors used in scientific descriptions of research. Gluck also points to the division of labor in a modern laboratory, which diffuses responsibility for individual animals among a number of researchers and students. Rarely is a single animal the responsi- bility of a single person; more likely, the animal is simply one of many sources of data for the next publication. As we saw in Chapter 5, the growth in size of the laboratory from the small basement room of Magendie to the massive, quasi-industrial operation of Ehrlich had as one effect the distancing of the researcher from the research subject. Sociologist Arnold Arluke has further noted, in his 1992 essay "Trapped in a Guilt Cage," that uneasiness or guilt feelings among laboratory workers are simply acknowledged, and describes coping mechanisms that lab technicians and caretakers have developed in their day-to-day work with experimental animals.

One argument that has been advanced against Harlow's research is that it did not provide a good model for human behavior. Monkeys are not humans, and different species of monkeys act differently. While rhesus macaques react radically to isolation, other monkey species, even other macaques, show less severe behavioral changes. Maternal deprivation is important to rhesus macaques, but in other species deprivation of the father is more damaging. This criticism strikes at the heart of the research enterprise, and may never be fully answered. In psychology, as in Claude Bernard's quest for the laws of general physiology, the discovery of universals remains elusive. For every animal model shown to be inadequate, there are others that work very well indeed.

The case of thalidomide is often presented as a classic example of an inadequate animal model. Thalidomide was a sedative that began to be marketed in Europe in 1959, after animal testing on rats and rabbits had shown no ill effects. The dangers to fetuses of mothers' ingestion of drugs (or alcohol, or nicotine) were not well understood in the 1950s. But when taken by pregnant human females, thalidomide caused extensive and severe birth defects. It causes the same defects in other primates. So was this a question of the wrong animal model or of not enough testing? The U.S. Food and Drug Administration in 1959 refused to approve the marketing of thalidomide in the United States, citing inconsistent test results and requesting more testing. The few American women who had the opportunity to take it that year were given it by their physicians as an experimental drug. In a further twist, thalidomide recently has been shown to be effective in a wide variety of disorders, including skin diseases, some immunologic disorders, and even some forms of cancer.

Research on primates has served to show how very close to humans they are. Studies in the last thirty years on primate intelligence and behavior have revealed complex and highly intelligent creatures with intricate social organization, tool use, and even speech, all the salient characteristics that had been said to distinguish humans from beasts since antiquity. Jane Goodall (b. 1934) began her studies on chimpanzees in the wild in 1960, and over a thirty-year span traced the lives of three generations of chimpanzees at Gombe National Park in Tanzania. She discovered a complex social hierarchy and tool use, as well as meat eating (chimpanzees had been thought to be herbivores), warfare, and cannibalism. She even observed a polio epidemic. Goodall's focus on each animal as an individual personality—she named each chimpanzee—revolutionized primate research and emphasized the similarity of chimpanzees to humans. It countered the long-standing scientific tradition of impersonality and objectivity; a journal editor of an early article of Goodall's demanded that she change "he" and "she" to "it" when referring to chimpanzees. Subsequent research by Dian Fossey (1932–1985) on Rwandan gorillas and Biruté Galdikas (b. 1946) on orangutans in Borneo supported Goodall's views of the intelligence and complexity of the great apes.

Chimpanzees also could learn language. Beatrice and R. Allan Gardner began to teach sign language to Washoe, a young chimpanzee, in the late 1960s. Roger Fouts took up this program in the early 1970s. Skeptics claimed that the animals were simply imitating their keepers, but Fouts's Washoe soon displayed an extensive vocabulary and the ability to make sentences, and taught her stepson, Loulis, to sign as well. Duane Rumbaugh and Sue Savage-Rumbaugh used a keyboard with symbols rather than sign language, with similarly impressive results. Savage-Rumbaugh later taught two chimpanzees to play computer games, and they proved as skilled (and addicted) as any nine-year-old boy. Rumbaugh explored chimpanzees' ability to learn number relationships.

Some find this evidence of chimpanzee intelligence exhilarating. Others find it depressing. If chimpanzees are so smart, and so closely related to humans, should they be confined in zoos and used as experimental animals? Although the number of chimpanzees used in biomedical research is small compared to some monkey species, their use in AIDS research makes them highly visible. This visibility has led in some cases to improved treatment for experimental animals. Goodall's advocacy, for example, led to improved conditions for chimpanzees in AIDS research at a National Institutes of Health–funded facility in Maryland.

In the mid-1990s, a group of animal rights activists, researchers (including Goodall), and others conceived of The Great Ape Project. The object of the project is, in its words, "to include the non-human great apes [chimpanzees, gorillas, and orangutans] within the community of equals by granting them the basic moral and legal protection that only human beings currently enjoy."* By including apes in the category of "persons" rather than property, great apes would be guaranteed the basic human rights to life, individual liberty, and freedom from torture. While the organizers acknowledge that many humans do not now possess these rights, they believe that this is not in itself a reason to withhold them from apes, who, they argue, are worthy of them from both scientific and ethical points of view. In practical terms, this would mean an end to the use of apes in zoos, entertainment, and research. The ultimate goal of

---

* "The Great Ape Project," www.greatapeproject.org, accessed 24 October 1999.

the project is a declaration of the rights of great apes by the United Nations, which would lead to the establishment of protected territories for the apes to live out their lives. In his book *Rattling the Cage* (2000), lawyer Steven Wise argues that apes have standing under the law.

The Great Ape Project is about not only the use of animals in research but also their conservation and survival in the wild. These two issues are closely related, as the decimation of the population of rhesus macaques in the 1950s demonstrated. Whether one views the project as a pipe dream of animal rights advocates or as a reasonable proposal, it forces us to confront our paradoxical relationship with other primates.

Experimentation on nonhuman primates greatly increased in the twentieth century. In areas such as polio research (and more recently, AIDS research), researchers argued that primates were essential models for certain human diseases. Their similarities to humans in physical and emotional characteristics made them excellent subjects in the burgeoning field of psychological research. But these very similarities made the use of primates ethically problematic. As evidence mounted of the intelligence, language use, and social structures of primates, questions also mounted concerning the ethics of primate use. At the same time, the development of the clinical trial raised questions about experimentation on humans. Are these questions interrelated?

Conclusion

Human Rights, Animal Rights, and the Conduct of Science

The second half of the twentieth century saw a strong movement toward public regulation of experiments on human beings and animals. In the case of human experimentation, initial impetus came from the exposure of atrocities performed in the name of science on the inmates of Nazi concentration camps. The trial of doctors before the Nuremberg War Crimes Tribunal in 1946 impelled the tribunal to lay down a code of conduct designed to protect the rights of human subjects. The first prerequisite was the informed consent of the subject, which meant that he or she must be mentally competent, uncoerced, and fully aware of the possible consequences of the experiment. The goal of the experiment must be important enough to justify the risk to the subject and must be unattainable by other means. The experiment must be conducted by qualified persons and must be terminated whenever the investigator judged that its continuance would result in death or permanent harm to the subject. It must also be terminated if and when the subject desired it.

The Nuremberg Code established a regime that was to be revised and supplemented by a succession of international declarations under the auspices of the World Medical Association, starting with the Helsinki Declaration of 1964. In the 1960s, however, it came to public attention that medical researchers in the United States and the United Kingdom violated the rights of human subjects routinely and with impunity. This scandal demonstrated the need for legislative regulation by national governments. At the same time, the growing interest in mankind's relationship to nature and the environment created a climate of opinion favorable to similar regulation of animal experimentation.

Research on Human Subjects and the Rise of Regulation

The concentration camp experiments were an extreme expression of the values of a profession that was deeply complicit in Nazi racism. Medical scientists took the lead in elaborating the distinctively German strain of eugenics called *Rassenhygiene* (racial hygiene), which classified certain ethnic groups, along with sexual deviants and the mentally ill, as a threat to the nation's health. Medical personnel took the lead in implementing the Nazi solution to the problem, a solution that embraced sterilization, euthanasia, and ultimately mass murder. It was logical for Nazi scientists to view these "defective" specimens of humanity as fit material for medical experiments for the good of the Reich. The resulting research ranged from the seemingly banal to the obviously horrendous. They included studies of the effects of high altitude, hypothermia, and long-term starvation; of the efficacy of sulfanilamide in treating gunshot wounds; and the effectiveness of drugs and other treatments. Men had their genitals irradiated and were then castrated. Children were decapitated in order to study brain development at different stages of growth. Methods of sterilization and euthanasia were tested, and prisoners were used as subjects to train surgical students. At Auschwitz the notorious Josef Mengele experimented on fifteen hundred sets of twins, many of them children. Over one thousand people died as a direct result of such experiments, and many more were permanently injured, both physically and psychologically.

It would be easy to condemn these experiments—most of them conducted without anesthesia and in horrific circumstances—as the work of madmen, but bioethicist Arthur Caplan warns that to do so would be to deny their character as a logical expression of the values of German medical science. In his view, "mainstream biomedicine in Germany boarded the Nazi bandwagon early, stayed on for the duration of the Nazi regime, and suffered few public second thoughts or doubts about the association even after the collapse of the Reich."* Certainly, none of the doctors tried at Nuremberg pleaded insanity; rather, they defended their actions as consistent with the values of science and their duties as scientists. They claimed that their subjects were volunteers who had been promised freedom in return for their collaboration, or that they had experimented only on people who were doomed to die in any case, or that the experiments gave their victims a way to expiate their crimes. They argued that they were following orders, and that their training as scientists gave them no grounding in ethics that might justify refusing those orders. They asserted that they had acted in defense of a state engaged in total war, and that it was legitimate that a few should have been made to suffer for the good of the many.

The Nuremberg Code was designed to test these pleadings in the light of established human rights and standards of human experimentation. It is therefore important to recognize that, while the Nazi experiments may have plumbed unparalleled depths of barbarity, they differed only in degree from

---

*Arthur L. Caplan, "How Did Medicine Go So Wrong?" in Arthur L. Caplan, ed., *When Medicine Went Mad: Bioethics and the Holocaust* (Totowa, N.J.: Humana Press, 1992), 57. My discussion of Nazi medical science is based on the essays in this book.

experiments that were then being conducted in the United States, and that continued to be undertaken there and in other democratic countries. The most notorious example is the so-called Tuskegee Syphilis Study, which was conducted from 1932 to 1972 under the aegis of the U.S. Public Health Service. In this experiment, four hundred black males, all suffering from syphilis, were left untreated for decades in order to study the natural history of the untreated disease. They were not told of their condition or of their status as experimental subjects and were duped into believing that they were receiving special medical care from Public Health Service physicians. On the contrary, their malady remained untreated even after it became known, early in the 1940s, that penicillin was an effective cure in the early stages of the disease.*

The Tuskegee experiment was not revealed to the public until 1971, but during the 1960s other evidence demonstrated the extent to which American medical researchers felt themselves exempt from the Nuremberg standards. In 1965 Henry K. Beecher, the Dorr Professor of Anesthesiology at Harvard University, alerted the national press to a number of unethical studies of which he was aware. He had earlier raised these concerns in a professional forum and now "went public" only because he was outraged by his colleagues' indifference to the issue. After some tribulations, including rejection by the *Journal of the American Medical Association,* he was able to publish a short account of twenty-two of these experiments in the *New England Journal of Medicine.* One was a study of typhoid fever that entailed withholding from a control group a remedy that was known to be effective. This group sustained twenty-three deaths, which could have been prevented with the appropriate medication.†

While Beecher was pursuing the issue in the United States, a physician, Maurice Pappworth, was doing so in Britain. He too encountered indifference and hostility in professional circles, which goaded him into public revelation. In his book *Human Guinea Pigs,* he suggested outright that there was no essential difference between the practices exposed by Beecher and himself and those for which the Nazi doctors had been condemned. Indeed, one striking feature of these experiments was that many of them were inflicted on individuals whom Nazi ideology might have branded as defectives. The typhoid study was undertaken on charity patients. Other experiments exposed by Beecher were

---

* Jay Katz, "Abuse of Human Beings for the Sake of Science," in Caplan, ed., *When Medicine Went Mad,* 248–50.

† Henry K. Beecher, "Ethics and Clinical Research," *New England Journal of Medicine* 274 (1966): 1354–60. This article is reprinted with a historical introduction in Jon Harkness, Susan E. Lederer, and Daniel Wikler, "Laying Ethical Foundations for Clinical Research," *Bulletin of the World Health Organization* 79 (2001): 365–72.

perpetrated on senile hospital patients and on "mental defectives or juvenile delinquents who were inmates of a children's center."* One study, which continued for several years after he reported it, took place at Willowbrook State School for the Retarded on Staten Island. This experiment, which was approved and funded by the Armed Forces Epidemiological Board, involved deliberately infecting inmates with a mild form of hepatitis. And, of course, there were the four hundred black victims of the Tuskegee study, which had not yet come to light.

The most common breach of the Nuremberg Code was the failure to obtain informed consent. Typically, the subjects of these experiments were given inadequate information or none at all. In other cases they were not in a position to give uncoerced consent as required by the code. Pappworth raised the problem of experiments on prisoners, which had a history dating back beyond eighteenth-century Britain to ancient Persia and Egypt. Noting that the Nazi doctors had cited contemporary experiments by Americans and others in their defense, he questioned whether prisoners serving long sentences, or confined in appalling conditions, were in a position to give uncoerced consent when offered relief from their plight in return for serving as human subjects. One of Beecher's examples involved a different sort of coercion: exploitation of a mother's grief. "Melanoma was transplanted from a dying daughter to her volunteering and informed mother, 'in the hope of gaining a little better understanding of cancer immunity and in the hope that the production of tumor antibodies might be helpful in the treatment of the cancer patient.' Since the daughter died on the day after the transplantation of the tumor into her mother, the hope expressed seems to have been more theoretical than practical." The mother died of metastatic melanoma some fifteen months later.†

Many of the ethical breaches recorded above were also committed by scientists who conducted radiation experiments on behalf of the United States government during the Cold War. Two well-documented instances concern the administration of tiny doses of radioactive material to children at two Massachusetts facilities for mentally retarded children. In the case of one experiment, conducted between 1950 and 1953 at the Walter E. Fernald School, two letters were used at different times to request parental permission for experiments on the children. Neither letter mentioned that the children would be exposed to radiation, and both falsely suggested that the experiments would benefit the

---

* M. H. Pappworth, *Human Guinea Pigs: Experimentation on Man* (London: Routledge and Kegan Paul, 1967), 185–87.

† Pappworth, *Human Guinea Pigs*, 60–68; Beecher, "Ethics and Clinical Research," 1359.

subjects. An investigation more than forty years later concluded that in this case at least, the children had probably suffered no harm. Nevertheless, the Massachusetts Institute of Technology and the Quaker Oats Company (which had collaborated in doctoring the children's breakfasts) settled a lawsuit out of court for $1.85 million.*

By the 1970s Beecher's and Pappworth's exposés, and the revelation of the Tuskegee study, had demonstrated that biomedical researchers could not be trusted to adhere to the principles of the Nuremberg Code without the coercive inducement of national regulation. In the United States, this took the form of the National Research Act of 1974, which established the regulatory apparatus that still prevails. It included a formal requirement of written consent for experimentation on human subjects and the establishment of institutional review boards (IRBs) to evaluate proposed experiments. The act also set up a national commission to identify the basic ethical principles that should underlie the conduct of research on human subjects. The result was the Belmont Report of 1979, which laid down three guiding principles for research on human subjects: respect for persons, which required the fully informed consent of the subject; beneficence, defined as an obligation to maximize possible benefits and minimize harm to the subject (as opposed to an earlier emphasis on simply avoiding harm); and justice, defined as a fair distribution of the benefits and burdens of research. The principle of justice dictates that vulnerable groups should not be used as subjects, but it has also been invoked to require that particular groups be included in a research design. In the early 1990s, the definition of a "normal" research subject as a white male was drastically revised to include women and, in some instances, ethnic minorities. The 1993 National Institutes of Health (NIH) reauthorization bill required that both female and male subjects be used at all stages of clinical research.

Regulation of Research on Animals

Despite having revulsion against Nazi excesses to stimulate national regulation of research practice, three decades would pass before the concerns raised by unethical experimentation on humans resulted in such regulation in the United States. Lacking such a stimulus, the movement to regulate research on animals proceeded even more slowly. In the early twentieth century, antivivisectionists

---

*The radiation experiments are documented in United States, Advisory Committee on Human Radiation Experiments, *Final Report* (Washington, D.C.: U.S. Government Printing Office, 1995). See also Zareena Hussain, "MIT to Pay Victims $1.85 Million in Fernald Radiation Experiment," *The Tech* (Cambridge, Mass., 7 January 1998).

had made explicit connections between experimentation on animals and humans: George Bernard Shaw, for instance, argued that doctors who experimented on animals would be more likely to experiment on their patients. The Nazi example challenged this connection, however (Nazi ideology, with its reverence for nature, was strongly antivivisectionist, though animal experimentation did not cease under the Reich), and after World War II the impulse to control human experimentation worked against the control of animal experimentation. The Nuremberg Code precluded research on human subjects in pursuit of ends that could be attained by other means, and it required prior experimentation on animals as a means of reducing the risks to the human subject. Apart from these considerations, as we have seen, the advances in medical science in the years before World War II helped to defuse antivivisectionist criticisms of research, and the increased popular respect for the medical profession effectively challenged the old image of the physician-vivisector.

By 1960, as government support for scientific research reached unprecedented heights, the antivivisection movement seemed to be at its nadir. But the intellectual and social turmoil of the 1960s revived and transformed the old antivivisection debates. New theories of animal rights and animal liberation, analogous to theories relating to human beings, stoked the fires of antivivisectionist protest and produced profound changes in scientific practice. Perhaps the collision could not have been avoided, as the increasing scale of research on animal and human subjects by the 1960s coincided with increased consciousness of human and animal rights.

As in the nineteenth-century movement against vivisection, the new animal rights movement centered in English-speaking countries, particularly Britain and the United States.* A founding document was the 1975 book *Animal Liberation* by an Australian philosopher, Peter Singer, now a professor at Princeton University. Singer collected information on the exploitation of animals, both in science and in the food industry, which had appeared in scattered, often obscure sources. He framed this information within a coherent moral philosophy, which situated the concept of animal rights within the expanding contemporary perception of human rights. His book was published under the imprint of the *New York Review of Books,* a respected literary and political weekly that was (and is) widely read by the educated public.

Animal Liberation reached a far wider audience than earlier works on animal exploitation, and it created a sensation. It presented animal liberation as "the obvious next step" to the liberation movements of the 1960s, such as

---

* "Animal rights" describes only one philosophical strand of this movement. For the sake of brevity I sometimes use it to refer to the various modern animal protection movements.

women's liberation and opposition to racism.* Singer employed the term *speciesism* (coined by animal activist Richard Ryder a few years earlier) to convey the parallels with racism and sexism. He barely mentioned the antivivisection movement and accused traditional animal welfare organizations, such as the RSPCA and the American Society for the Prevention of Cruelty to Animals (ASPCA), of providing a cloak for the continuing exploitation of animals. Animal liberation was something new.

In at least one respect, however, Singer's approach did resemble that of the old antivivisectionists: for him, as for them, the issue was one of pain. Specifically, he based his philosophical stance not on the idea of rights but on Jeremy Bentham's philosophy of utilitarianism, discussed in Chapter 3. This philosophy is based on the notion of interests: that the true measure of moral principle is the capacity for experiencing pain or pleasure, not some abstract notion of rights. Action—of governments and individuals—should be based on maximizing happiness of the greatest number of individuals. Because animals too experience pain and pleasure, their interests have equal status with those of humans. They are our moral equals. Therefore, in Singer's view, if we are to treat animals in accordance with our moral behavior toward humans, we should maximize pleasure and minimize pain for them as well.

Animal liberation quickly became a movement, which attracted many younger people who would never have gone near the mainstream animal welfare organizations. The newer groups were considerably more radical and activist, and by the late 1970s protests against animal use in experimentation, product testing, the food industry, and the fur industry were widespread. The protesters employed the full arsenal of modern communications media in their cause, resulting in greatly increased public consciousness of animal issues. Groups such as People for the Ethical Treatment of Animals (PETA) have lobbied for the end of cosmetics testing on animals and have revealed serious abuses in research laboratories. Particularly in the latter case, however, their "undercover" methods have sometimes been questioned, and some radical groups have destroyed laboratories, "liberated" research animals, and harassed scientists to the extent of issuing death threats in the name of animal rights. Singer, among others, has decried the use of violence, but it remains a prominent part of the animal protection movement for groups such as the Animal Liberation Front, who declare that any strategy is justified. This attitude has led to widespread polarization between researchers and activists.

By the 1980s, the use of animals in commerce and research was a hot topic,

* Peter Singer, *Animal Liberation*, 2d ed. (1990; reprint, New York: Avon Books, 1991), "Preface to the New Edition," viii.

covered in dozens of books and articles. Not all of them accepted Singer's utilitarian philosophy as a basis for the moral treatment of animals. The American philosopher Tom Regan offered an alternative in *The Case for Animal Rights* (1983). Regan argued that mammals at least, though perhaps not all animals, possess attributes that give them the same inherent value, and thus the same rights, as human beings. These attributes include beliefs and desires, memory, and an emotional life that includes feeling pain and pleasure. Others have turned this "quality of life" argument against Regan, however, by arguing that human beings have a greater quality of life, and thus an inherently greater value, than any animal. From another perspective, R. G. Frey has asserted that the quality-of-life argument can apply only to fully functional humans. He proposes that brain-damaged humans have a lesser quality of life than healthy animals and may justifiably be used as research subjects.

In *The Rights of Nature* (1989), environmental historian Roderick Nash implicitly endorsed Singer's view of the inevitability of animal liberation by expounding his own concept of an expanding circle of rights. Nash observes that notions of who, or what, possesses the fundamental rights of existence have changed over time to encompass wider and wider circles of human society and finally the nonhuman world. From the idea that only kings had rights, or only white men, the circle has gradually grown to include nonwhite men, women, and animals, and finally will include all of nature. In *The Sexual Politics of Meat* (1990), Carol J. Adams added a feminist dimension to these arguments, echoing some nineteenth-century antivivisectionists who argued that women and animals were both exploited. Adams argues that "meat is a symbol for . . . patriarchal control of animals." In rejecting the male culture of violence that surrounds meat eating, Adams finds common ground between feminists and animal activists.*

Nash sees his ideas as a distinctively American (or maybe Anglo-American) expression of liberal doctrines, and indeed it is especially applicable to the United States, which has the idea of fundamental rights embedded in its origins. Nash and Singer share a progressivist view of history: that we are better, and more enlightened, than our past. But in the history of many other nations, rights have neither been inevitable nor linear in their development or application. Nor is the relationship between animal rights and environmental ethics a straightforward one. Philosophers of the biocentric school, such as J. Baird Callicott, hold that animal liberation and environmental ethics are profoundly

---

* Carol J. Adams, *The Sexual Politics of Meat* (1990, reprint, New York: Continuum, 1995), 16.

opposing philosophies. Both the utilitarian theory championed by Singer and the rights theory of Regan apply to animals who can feel pain. But the question of pain, argued Callicott in 1980, is irrelevant to the ecosystem, which includes pain, death, and violence as part of the natural cycle of life. An environmental ethic concerns the good of the biotic community as a whole, and its interests supersede the interests of individual members. Callicott offered this "holistic" philosophy as an alternative to the individualism of the rights/liberation philosophies.

In *Animals and Why They Matter* (1983), Mary Midgley summarized several philosophical arguments concerning animals and concluded that animals of course matter. Without espousing any particular position, she concluded, with a welcome open-mindedness, that humans should embrace the whole of nature. Another philosopher, Richard Sorabji, emphasizes the relationships between animals and people—which differ substantially in different cultures—as a basis of discussion. For both, the questions are too complex to reduce to slogans.

The climate of opinion that produced these statements was conducive to more stringent regulation of animal experiments, but actual regulation has been strongly influenced by a more pragmatic work: *The Principles of Humane Experimental Technique* (1959) by W. L. M. Russell and R. L. Burch of the Universities Fund for Animal Welfare, a British organization.* Russell and Burch propounded what they called the "three Rs": Replacement, Reduction, and Refinement. "Replacement" echoed the Nuremberg Code in requiring that animal experimentation be used only when no other means would achieve the goal of the research. Their suggested alternatives included using microorganisms or tissue cultures. "Reduction" meant carefully designing experiments, avoiding the trial-and-error approach as far as possible, and employing various techniques to minimize variability and thus reduce the number of subjects needed. "Refinement" applied above all to behavioral studies, and entailed the use of lower rather than higher vertebrates and, in general, taking care to choose the appropriate animal for the study.

The influence of the "three Rs" is palpable in the 1985 Animal Welfare Act, which currently governs animal experimentation in the United States. This legislation revised a 1966 act that laid down minimal rules for animal welfare, to be enforced by regular inspection by the federal Department of Agriculture. The 1985 act established much higher standards of care and inspection

---

*W. L. M. Russell and R. L. Burch, *The Principles of Humane Experimental Technique* (London: Methuen, 1959).

and required the setting up of institutional animal care and use committees (IACUCs), the counterpart of IRBs for human subject research. It imposed a duty to consider the use of alternatives to animal experimentation and set up a central clearing house for information on alternatives.

In most Western countries today, laws of varying effectiveness and rigor regulate animal experimentation. Although they differ widely in detail, all have the same aims, as Judith Hampson, an expert on animal welfare issues, outlines them:

- to define legitimate purposes for which laboratory animals may be used
- to exert control over allowable levels of pain or other distress
- to provide for inspection of facilities and procedures
- to ensure humane standards of animal husbandry and care
- to ensure public accountability

Hampson argues that no current system has successfully achieved all of these goals but that the main elements of an effective system—legislation, a review apparatus, and rules for care and use—exist in many countries.* Differing balances among these general aims are an expression of national values and the specific historical, social, and political circumstances that have engendered them.

Current Problems and Issues

In the 1970s and 1980s, a comprehensive national regulatory apparatus to govern the use of human and animal subjects in scientific research was instituted in the United States. Most European countries have also adopted some form of legislative control over such activity. The European Union adopted a resolution in 1986 requiring member nations to regulate animal research, and subsequent legislation by EU states must conform to the standards of the resolution. In general, European legislation is more stringent than in the United States in its restrictions on animal use. Although there is wide variation among EU nations, the emphasis is on national rather than local enforcement. The American approach, in contrast, emphasizes institutional rather than national decision making and evaluation. In Britain, for example, animal research is regulated by the home secretary, a minister of the central government, and inspectors under the secretary's supervision assess and license research projects. Although there are local animal ethics committees, their role is purely

---

*Judith Hampson, "Legislation and the Changing Consensus," in Gill Langley, ed., *Animal Experimentation: The Consensus Changes* (New York: Chapman and Hall, 1989), 220.

advisory.* In the United States in particular, some spectacular lapses in the regulatory system, together with rapid advances in biotechnology, have combined to raise troubling questions about the ethical limits of biomedical research and the capacity of regulatory institutions to impose limits on scientific curiosity and its commercial exploitation.

In the United States, most regulation takes place at the level of the individual institution, which could be a university, private company, or federal laboratory. Committees (either IACUCs or IRBs) composed of scientists and nonscientists review protocols for their adherence to the specified standards. They have the power to stop research projects and to require changes in protocols designed to ensure compliance with those standards. However, as Andrea L. Beach and David E. Wright remark with respect to IRBs, despite the increase in regulations and federal supervision, this remains largely an honor system,† and its effectiveness varies from one institution to another. At some universities, the committee meets at least monthly, every member has had an opportunity to review the protocols, and there is serious discussion. In other institutions the committee serves as little more than a rubber stamp. Ethical considerations are not within their purview (as they are in Australia, for example). The committee's purpose is to ensure compliance with the laws governing research and so to protect the institution, ensuring that external agencies continue to fund research.

For a conscientious committee, protocol review means a lot of work. The protocols may vary from a few to more than twenty pages and, at a large institution, the committee may have to review more than fifty a month. The terminology can be daunting even for scientists outside the immediate subdiscipline of the protocol; for nonscientist members it can be incomprehensible, even though the protocol is supposed to be written for a nonexpert. In these cases, committee members can consult with the researcher and send back the protocol for rewriting. Mastering the tangle of laws, regulatory agencies, and acronyms can be another lengthy task. In the case of animal experimentation, for instance, institutions receiving federal funding (which includes virtually all American universities) are also subject to rules issued by the National Institutes of Health and administered by the Public Health Service. Unlike the 1985

---

*European legislation is summarized in European Science Foundation Policy Briefing 15, "Use of Animals in Research," www.esf.org/policy/sci_pol_pubs.htm. Regulation in the U.S. and U.K. is compared in M. Allen, "US and UK Control of Lab Animal Experimentation," *Lab Animal* 27 (May 1998): 34–39.

†Andrea L. Beach and David E. Wright, "A Short History of Regulations Protecting Human Subjects Research," *Research Integrity* 4:1a (spring 2000): 5.

Animal Welfare Act, which excluded rats, mice, and birds, NIH policy includes rats and mice (the exclusion of rats and mice from the act has been challenged in court; the case is ongoing). In addition, the *Guide for the Care and Use of Laboratory Animals,* developed in 1963 by laboratory veterinarians as a set of self-regulatory guidelines, provides another set of rules. The *Guide,* published by the National Academy of Sciences and now in its seventh edition, presents a set of standards that is generally, but not always, in keeping with federal policy. Voluntary groups such as Public Responsibility in Medicine and Research (PRIM&R) and Applied Research Ethics National Association (ARENA) hold regular meetings, attended by hundreds of administrators, scientists, and committee members, to provide guidance through the regulatory process.

Regulation is thus necessary but often difficult. Scientists complain of the regulatory burdens placed on experimentation and, perhaps with more justification, of the enormous burden of paperwork required to comply with those regulations. Committee members complain about poorly written protocols and unclear rules. Perhaps few in the United States would care to move to a more centralized system on the European model, but there is room for improvement in the United States system. The inspection apparatus is inadequate and the tangle of agencies inefficient. Much depends—some would say too much—on the integrity of individual scientists and committee members. The committee's duty is to review the experimental procedures, as set forth in the protocol, to ensure conformity to the law. But the investigator then returns to the laboratory, without the committee looking over his or her shoulder to ensure that the research conforms to the protocol.

In the field of human experimentation, two recent tragedies have highlighted the failings of the system. The first occurred in September 1999, when eighteen-year-old Jesse Gelsinger died in a gene therapy trial at the University of Pennsylvania. The resulting investigation revealed serious lapses, including ludicrously inadequate record keeping by the researchers. Deaths of experimental animals went unreported, as did serious human side effects. Changes in the experimental design, which should by law have led to the preparation of a revised protocol for review, were also not reported. Most seriously, the researchers were charged with neglecting informed-consent procedures. Jesse Gelsinger was not informed of the death of experimental monkeys from the procedure he was about to undergo. Because of his poor health, he was not actually eligible to participate in the program. The federal government shut down the Penn program—the largest in the country—in January 2000, a move that shocked the public and the research community. Subsequently it was allowed to resume but was barred from working with human subjects.

Then, in June 2001, twenty-four-year-old Ellen Roche, a healthy subject, died in an asthma experiment at the Johns Hopkins University. A drug was administered to stress (in research jargon, to "challenge") her lungs. It proved only too effective. An internal investigation found several lapses in the design, review, and conduct of the experiment. The IRB should never have approved it because the researcher did not present sufficient evidence that the drug was safe. The consent form described the drug as a medication; it did not disclose that it was no longer in clinical use, was not approved by the U.S. Food and Drug Administration, and could cause severe side effects, including death. The study had not been suspended, as it should have been, when the drug induced a nine-day cough in an earlier subject. Faulted by the FDA for not obtaining its approval for the experiment, the researcher blamed the IRB, on whose guidance he relied, for not telling him to do so.

Shortly afterward, the federal Office for Human Research Protection briefly suspended nearly all (some 2,400) human medical experiments at Johns Hopkins. It alleged widespread lapses in safety procedures, including IRBs whose workload prevented proper review of research proposals or supervision of research in progress. Federal officials faulted the complex terminology of some consent forms and found cases where the risks of the experiment, as stated in the form, differed from those communicated to the review board. Other ethical lapses concerned the issue of uncoerced consent. Roche was a technician in the laboratory where the experiment was conducted. It was not suggested that she had been coerced into taking part, but that her trust in the research faculty might have unduly influenced her consent. Another element of undue consent was the significant honorarium ($365) promised for her participation. These were issues that had been reviewed by Maurice Pappworth decades earlier.*

Recent rapid advances in biotechnology aggravate the problem of regulation. New discoveries and techniques have complicated the ethical picture and rendered regulatory tasks even more knotty. The old rules do not cover activities such as cloning and gene therapy. The development of transgenic mice in the 1970s—mice with new genetic information introduced into a genome—transformed biological research in several ways. Custom-made research animals, such as mice that are genetically obese, or epileptic, or prone to cardiac

---

* This account is based on reports in the Baltimore *Sun*, July 17, 20, and 23, 2001, and an article by Michael Susko, "Researchers Must Ensure Woman Didn't Die in Vain," Baltimore *Sun*, July 25, 2001, 13A. See also Pappworth, *Human Guinea Pigs*, 192, 216.

disease, may be viewed as a more efficient use of animals in research. But the ethical implications for the animal's autonomy have been neglected. Pig heart valves have long been used to repair human hearts, but soon there will be transgenic pigs with human organs that can be harvested for transplant, a process known as xenotransplantation. When surgeons transplanted a baboon's heart to Baby Fae in 1984, the furor that followed focused mainly on the baby, who died, although some consideration was given to the baboon, who, of course, also died.

The cloning of the sheep Dolly in 1996 revealed once more the inadequacy of current ethical systems to address new technologies, as commentary focused mainly on the possibility of cloning a human being rather than on the implications of the process itself. Most biologists viewed Dolly as an interesting, but hardly earth-shattering, example of reproductive research. Cloning is a difficult procedure—277 unsuccessful attempts were made on sheep before Dolly—and it has been used, with varying success, on a number of mammals to develop genetically engineered animals with specific traits, including the capacity to produce specific human hormones. The focus on the possibility of human cloning has obscured these goals and has led several countries and four states in the United States to ban any research on cloning. Recent hearings in the U.S. House of Representatives pitted scientists and bioethicists who are opposed to the idea of cloning against its advocates. Opponents argue that the procedure has shown itself in animals to be dangerous and unpredictable in its results; the success rate is very low, and some cloned animals have inexplicable defects. The National Bioethics Advisory Commission, convened in 1997 to consider the prospect of human cloning, similarly concluded that the procedure is not currently safe to use on human beings. The commission advised a temporary ban on human cloning research in the United States, which is still in effect. Intertwined with these discussions are questions about the use of human embryonic stem cells in research. Advocates of human cloning, including a medical team in Italy and a group that believes that life on earth was created through genetic engineering by extraterrestrials, believe that scientific obstacles can be overcome. But would cloning a human, if procedures existed, be right or wrong?

Gene therapy on human subjects is another area of controversy. It has been reported as the wave of the future for medicine, but recent scandals have revealed many ethical and legal problems. Because of the explosive growth of the biotechnology industry, many scientists, even those affiliated with universities, have substantial commercial interests in their research. Although gene therapy is subject to regulatory oversight, adverse reactions have sometimes been reported tardily or not at all. While protocols may have met the standards of IRBs, subsequent laboratory practices may not have followed those standards. As noted earlier, boards do not have the power to force researchers to act ethically once they are back in the laboratory.

These complex ethical issues return attention to familiar philosophical debates about how we treat humans and other animals. The answer has everything to do with our view of the world as a whole. At this point in our history, it seems as though researchers and their opponents are operating from vastly different world-views. Protesters deride scientists as cruel, unthinking Cartesians, while scientists increasingly adopt a bunker mentality that dismisses any criticism as ill-informed. In fact, both sides need to acknowledge the difficult and ambiguous relationship between modern science and ethics. Even as recently as the 1960s, researchers operated according to what seemed to them to be a generally agreed upon set of ethical standards. They no longer have that luxury, and there are wide disagreements on ethical standards, in part because discoveries made in the past and in the present have repeatedly run up against those supposedly solid standards. Even if we do not agree with their views, Peter Singer and others have courageously tried to reinvent ethics to speak to the modern condition.

Historians believe that we must see the past in its own terms, avoiding judgment on the supposition that we are more enlightened or simply smarter than people in the past. The assumption that modern Westerners, and Americans in particular, are at some kind of culminating point in the history of ethical or scientific thinking is misleading and simply wrong. Recent biomedical science has created situations that our ethical systems have great difficulty in comprehending.

We may stand at a crossroads in terms of biomedical research and research ethics. We are finding out more about the human and animal body than we have ever known before, and we are not sure what to do with that knowledge, either practically or ethically. In the next decade, we will all have to make decisions about our lives, our privacy, and our ethical stance on certain critical issues. Research on reproduction, embryological development, and the genetic code will lead to an enormous number of techniques, therapies, and ethical questions. How will knowledge of the human genome be used? What is the future of gene therapy and genetic engineering? Will human cloning become a reality? Related to the last question is the use of human fetal tissue for research and therapy, a question that, especially in the United States, enters the murky and emotionally charged arena of debates about abortion. Is it ethical to breed animals to provide human body parts? Could it ever be ethical to breed humans for that purpose? Even if we know how, should we do it?

Another set of questions centers on emerging infectious diseases and the reemergence of those that were thought to be conquered. Research efforts must be redoubled to discover new drugs and vaccines, and this will undoubtedly use millions of animals. What moral principles can we develop to deal with such widely varying and troubling problems?

Where should we stand on these issues? Animal and human experimentation are necessary aspects of modern research and will not likely disappear in the foreseeable future. Regulation is necessary and in many cases should be strengthened. It is not up to scientists alone to determine what the values of our society are. It is up to all of us. But we cannot determine these values without being adequately informed about past and present ideas and practices. Scientists and nonscientists alike are beginning to discover that there are different, but not mutually exclusive, ways of looking at scientific and ethical problems. One hopes that this book will assist readers as they as citizens bring their own points of view to bear on these problems.

Suggested Further Reading

There are few general works on the topics of animal and human experimentation. *Vivisection in Historical Perspective,* ed. Nicolaas Rupke (New York: Routledge, 1987), is an excellent collection of articles that explore the topic from antiquity to the twentieth century, with the main focus on the nineteenth century. Individual articles are noted below. F. Barbara Orlans's *In the Name of Science* (New York: Oxford University Press, 1993) focuses on current practices but contains a short historical summary. Many antivivisectionist works contain historical material of varying value; some of these are listed below. General works in the history of biology also contain much relevant material on animal experimentation in particular; a good, if rather dry, text is T. S. Hall's *History of General Physiology,* 2 vols. (Chicago: University of Chicago Press, 1969); see also Karl Rothschuh, *History of Physiology,* trans. Günter Risse (New York: Krieger, 1973); Elizabeth Gasking, *The Rise of Experimental Biology* (New York: Random House, 1970). I know of no general history of human experimentation.

Preface

A convenient introduction to the works of Michel Foucault is *The Foucault Reader,* ed. Paul Rabinow (New York: Pantheon, 1984); see also his *The Order of Things* (1970; reprint, New York: Vintage, 1994). Some examples of studies of the body include Joan Cadden, *Meanings of Sex Difference in the Middle Ages: Medicine, Science, and Culture* (Cambridge: Cambridge University Press, 1993); Thomas Laqueur, *Making Sex: Body and Gender from the Greeks to Freud* (Cambridge: Harvard University Press, 1990); Thomas Laqueur and Catherine Gallagher, eds., *The Making of the Modern Body: Sexuality and Society in the Nineteenth Century* (Berkeley: University of California Press, 1987); Londa Schiebinger, *Nature's Body: Gender in the Making of Modern Science* (Boston: Beacon Press, 1993).

Innovative studies of the body from the perspective of anatomy include Jonathan Sawday, *The Body Emblazoned: Dissection and the Human Body in*

Renaissance Culture (New York: Routledge, 1995); Andrea Carlino, *Books of the Body: Anatomical Ritual and Renaissance Learning,* trans. John Tedeschi and Anne C. Tedeschi (Chicago: University of Chicago Press, 1999).

Feminist perspectives include the influential works of Donna Haraway, who discusses experimentation as part of a larger critique of science in general in *Primate Visions: Gender, Race, and Nature in the World of Modern Science* (New York: Routledge, 1989); *Simians, Cyborgs, and Women: The Reinvention of Nature* (New York: Routledge, 1991); *Modest_Witness@Second_Millennium .FemaleMan©_Meets_OncoMouse™: Feminism and Technoscience* (New York: Routledge, 1997). Carol J. Adams, in contrast, places experimentation within the context of ethical arguments against violence; see her *The Sexual Politics of Meat: A Feminist-Vegetarian Critical Theory* (New York: Continuum, 1990) and *Animals and Women: Feminist Theoretical Explorations,* which she co-edited with Josephine Donovan (Durham, N.C.: Duke University Press, 1995). Women as experimental subjects is among the topics in Helen B. Holmes and Laura M. Purdy, eds., *Feminist Perspectives in Medical Ethics* (Bloomington: Indiana University Press, 1992).

Among the many works employing sociological perspectives, especially influential has been Bruno Latour; see his *Laboratory Life: The Social Construction of Scientific Facts,* written with Steve Woolgar (Beverly Hills: Sage Publications, 1979). The concept of "boundary objects" was introduced in S. L. Star and J. R. Griesemer, "Institutional Ecology, 'Translations,' and Boundary Objects: Amateurs and Professionals in Berkeley's Museum of Vertebrate Zoology," *Social Studies of Science* 19 (1989): 387–420. Important collections include Adele Clarke and Joan Fujimura, eds., *The Right Tools for the Job: At Work in Twentieth-Century Life Sciences* (Princeton: Princeton University Press, 1992); Andrew Pickering, ed., *Science as Practice and Culture* (Chicago: University of Chicago Press, 1992); and Ilana Löwy and Jean-Paul Gaudillière, eds., *The Invisible Industrialist: Manufactures and the Production of Scientific Knowledge* (New York: St. Martin's Press, 1998). The *Journal of the History of Biology* featured a special section titled "The Right Organism for the Job" in vol. 26, no. 2 (1993): 233–367.

1 Bodies of Evidence

Richard Sorabji's very fine *Animal Minds and Human Morals* (Ithaca: Cornell University Press, 1993) explores ancient ideas about animals. John Passmore's "The Treatment of Animals," *Journal of the History of Ideas* 36 (1975): 195–218, is a useful short survey of ancient ideas. The Alexandrians receive a thorough treatment in Heinrich von Staden, *Herophilus* (Cambridge: Cambridge University Press, 1989); David C. Lindberg's *The Beginnings of Western Science* (Chicago: University of Chicago Press, 1992) offers much relevant material. Essential to understanding Aristotle is G. E. R. Lloyd, *Aristotle* (Cambridge: Cambridge University Press, 1968); for Galen see Owsei Temkin, *Galenism* (Ithaca: Cornell University Press, 1973). Nancy Siraisi's *Medieval and Renaissance Medicine* (Chicago: University of Chicago Press, 1990) is an excellent survey of ideas and practices with much material about anatomy. Relevant to chapters 1–3 is A. H. Maehle and U. Tröhler, "Animal Experimentation from Antiquity to the End of the Eighteenth Century," in the Rupke volume.

2 Animals, Machines, and Morals

Gweneth Whitteridge's *William Harvey and the Circulation of the Blood* (London: Macdonald, 1971) offers a thorough account of Harvey's discovery; see also William Harvey, *The Circulation of the Blood and Other Writings,* trans. K. Franklin, intro. Andrew Wear (London: Dent, Everyman's Library, 1990). Also important is Jerome Bylebyl, ed., *William Harvey and His Age: The Professional and Social Context of the Discovery of the Circulation* (Baltimore: Johns Hopkins University Press, 1979). The Royal College of Physicians of London produced an excellent video on Harvey in the late 1960s that is worth watching if you come across it. On Cartesianism and animals, the major source remains Leonora Cohen Rosenfield, *From Beast-Machine to Man-Machine* (1940, reprint, New York: Octagon,1968). On mechanical physiology, see R. G. Frank Jr., *Harvey and the Oxford Physiologists* (Berkeley: University of California Press, 1980), and T. M. Brown, *The Mechanical Philosophy and the "Animal Oeconomy"* (New York: Arno, 1981). Also useful is Anita Guerrini, "The Ethics of Animal Experimentation in Seventeenth-Century England," *Journal of the History of Ideas* 50 (1989): 391–407, and Marjorie Nicolson's *Pepys' Diary and the New Science* (Charlottesville: University of Virginia Press, 1965) offers an eyewitness account of Royal Society experiments. On Malpighi and his circle, see Domenico Bertoloni Meli, ed., *Marcello Malpighi, Physician and Anatomist* (Florence: L. Olschki, 1997).

3 Disrupting God's Plan

Essential to understanding the inoculation controversy is Genevieve Miller, *The Introduction of Smallpox Inoculation in England and France* (Philadelphia: University of Pennsylvania Press, 1957). A broader history of smallpox is Donald R. Hopkins, *Princes and Peasants: Smallpox in History* (Chicago: University of Chicago Press, 1983).

On Jenner, see Derrick Baxby, *Jenner's Smallpox Vaccine: The Riddle of Vac-*

cinia Virus and Its Origin (London: Heinemann Educational Books, 1981); on the modern eradication campaign, the main source is Frank Fenner et al., *Smallpox and Its Eradication* (Geneva: World Health Organization, 1988). Isobel Grundy's recent biography of Lady Mary Wortley Montagu contains much detail on her role in inoculation: *Lady Mary Wortley Montagu* (Oxford: Clarendon Press, 1999). The role of statistics is well explained in Andrea Rusnock, "The Weight of Evidence and the Burden of Authority: Case Histories, Medical Statistics and Smallpox Inoculation," in Roy Porter, ed., *Medicine in the Enlightenment* (Atlanta: Rodopi, 1995). On Hales, see D. G. C. Allan and R. E. Schofield, *Stephen Hales, Scientist and Philanthropist* (London: Scolar Press, 1980); a general account of eighteenth-century science is Thomas Hankins, *Science and the Enlightenment* (Cambridge: Cambridge University Press, 1985).

4 Cruelty and Kindness

The history of experimental physiology in the nineteenth century has a vast bibliography. John Lesch's *Science and Medicine in France: The Emergence of Experimental Physiology* (Cambridge: Harvard University Press, 1984) discusses the period before Bernard; F. L. Holmes's *Claude Bernard and Animal Chemistry* (Cambridge: Harvard University Press, 1974) closely studies the earlier years of Bernard's career. Also important is Joseph Schiller's *Physiology and Classification* (Paris: Maloine, 1980) and his "Claude Bernard and Vivisection," *Journal of the History of Medicine and Allied Sciences* 22 (1967): 246–60. The standard biography of Bernard remains J. M. D. and E. H. Olmsted, *Claude Bernard and the Experimental Method in Medicine* (New York: H. Schuman, 1952). For Marshall Hall, see Diana Manuel, *Marshall Hall (1790–1857): Science and Medicine in Early Victorian Society* (Amsterdam and Atlanta: Rodopi, 1996), and her article on Hall in the Rupke volume. The Anatomy Act of 1832 is the subject of Ruth Richardson's brilliant *Death, Dissection, and the Destitute* (London: Routledge, 1988). On William Beaumont see Ronald Numbers, "William Beaumont and the Ethics of Human Experimentation," *Journal of the History of Biology* 12 (1979): 113–35. The introduction of anesthesia is treated in Martin Pernick's *A Calculus of Suffering* (New York: Columbia University Press, 1985). Harriet Ritvo's entertaining *The Animal Estate: The English and Other Creatures in the Victorian Age* (Cambridge: Harvard University Press, 1987) looks at the relationships between animals and the Victorians; for the French, see Kathleen Kete, *The Beast in the Boudoir: Petkeeping in Nineteenth-Century Paris* (Berkeley: University of California Press, 1994), which covers more than its title indicates. The most important source for the nineteenth-century antivivisection movement remains Richard French, *Antivivisection and Medical Science in Victorian Society* (Princeton: Princeton University Press, 1975); see also James Turner, *Reckoning with the Beast: Animals, Pain, and Humanity in the Victorian Mind* (Baltimore: Johns Hopkins University Press, 1982), which is especially valuable for its analysis of American antivivisection. See also Lloyd Stevenson, "Science down the Drain," *Bulletin of the History of Medicine* 29 (1955): 1–26. Coral Lansbury, *The Old Brown Dog* (Madison: University of Wisconsin Press, 1986), and Lori Williamson, *Power and Protest: Frances Power Cobbe and Victorian Society* (London: Rivers Oram Press, 1998), discuss the role of women. Articles by Susan Lederer ("Controversy in America, 1880–1914") and Mary Ann Elston ("Women and Antivivisection in Victorian England") in the Rupke volume are also relevant. From the antivivisectionist point of view, see John Vyvyan's *In Pity and in Anger* (London: Michael Joseph, 1969) and *The Dark Face of Science* (London: Michael Joseph, 1971); Richard D. Ryder, *Victims of Science: The Use of Animals in Research* (London: Davis-Poynter, 1975).

5 The Microbe Hunters

C. E. A. Winslow's *The Conquest of Epidemic Disease* (Princeton: Princeton University Press, 1943) remains a classic account of the bacteriological revolution, especially useful for its long historical perspective. Paul de Kruif's *Microbe Hunters* (New York: Harcourt, Brace, 1926) conveys the excitement of research by one who participated in it; Sinclair Lewis's *Arrowsmith* (1925) is a fictional version of many of the same events. Both served as effective propaganda for science in the 1920s. Gerald Geison's controversial *The Private Science of Louis Pasteur* (Princeton: Princeton University Press, 1995), based on a close reading of his laboratory notebooks, reveals a less heroic but more human Pasteur; heroic stature is restored in Patrice Debré, *Louis Pasteur,* trans. E. Forster (Baltimore: Johns Hopkins University Press, 1998). For Koch, see Thomas Brock, *Robert Koch: A Life in Medicine and Bacteriology* (Madison, Wis.: Science Tech Publishers, 1988). Elie Metchnikoff, ed., *The Founders of Modern Medicine: Pasteur, Koch, Lister,* trans. D. Berger (New York: Walden, 1939), reprints original papers. Harry F. Dowling's *Fighting Infection: Conquests of the Twentieth Century* (Cambridge: Harvard University Press, 1977) is a reliable guide to the development of antibiotics. See also John Parascandola, ed., *The History of Antibiotics: A Symposium* (Madison, Wis.: American Institute for the History of Pharmacy, 1980); Harry M. Marks, *The Progress of Experiment: Science and Therapeutic Reform in the United States, 1900–1990* (New York: Cambridge University Press, 1997). Susan Lederer, "Political Animals:

The Shaping of Biomedical Research Literature in Twentieth-Century America," *Isis* 83 (1992): 61–79, discusses scientists' response to the perceived threat from antivivisectionists. On the experimental mouse, see Karen Rader, "The Multiple Meanings of Laboratory Animals: Standardizing Mice for American Cancer Research, 1910–1950," forthcoming in *Dead or Alive: Animals and Human Society* (Leiden: Brill).

6 Polio and Primates

A prize-winning account of current research on primates, Deborah Blum's *The Monkey Wars* (New York: Oxford University Press, 1994), offers a balanced view. A feminist-postmodern critique of primate research is offered in Donna Haraway, *Primate Visions*. The early years of polio research are detailed in Naomi Rogers, *Dirt and Disease: Polio before FDR* (New Brunswick, N.J.: Rutgers University Press, 1992). Jane S. Smith describes the development of the Salk vaccine in *Patenting the Sun: Polio and the Salk Vaccine* (New York: Morrow, 1990); for a scientist's view, see John R. Paul, *A History of Poliomyelitis* (New Haven: Yale University Press, 1971). Saul Benison, ed., *Tom Rivers: Reflections on a Life in Medicine and Science* (Cambridge: MIT Press, 1967), offers an insider's view of the hunt for a vaccine. On the 1954 field trials, see Allan M. Brandt, "Polio, Politics, Publicity, and Duplicity: Ethical Aspects in the Development of the Salk Vaccine," *International Journal of Health Services* 8 (1978): 257–70; Marcia Meldrum, "'A Calculated Risk': The Salk Polio Vaccine Field Trials of 1954," *British Medical Journal* 317 (1998): 1233–36. On the AIDS-polio link, see Edward Hooper's exhaustively researched *The River: A Journey to the Source of HIV and AIDS* (Boston: Little, Brown, 1999). For the latest information on this topic, see Brian Martin's excellent web site, www.uow.edu.au/arts/sts/bmartin/dissent/documents/AIDS/. Ann Giudici Fettner's *The Science of Viruses* (New York: McGraw-Hill, 1990) is an excellent introduction to the vocabulary and concepts of virology. Harry Harlow's famous presidential address to the American Psychological Association was published as "The Nature of Love," *American Psychologist* 13 (1958): 673–85; for an analysis of Harlow by one of his former students, see John P. Gluck, "Harry Harlow and Animal Research: Reflection on the Ethical Paradox," *Ethics and Behavior* 7:2 (1997): 149–61. Arnold Arluke summarized his work on animal workers in "Trapped in a Guilt Cage: How Do Scientists and Technicians Avoid Getting Close to the Animals They Work With?" *New Scientist* 134 (4 April 1992): 33–36. The Great Ape Project is described in Paola Cavalieri and Peter Singer, eds., *The Great Ape Project* (London: Fourth Estate, 1993); www.greatapeproject.org.

Conclusion

On the Nazi medical ideology, see Robert N. Proctor, *Racial Hygiene* (Cambridge: Harvard University Press, 1988). A recent survey of the ethical implications of Nazi medicine is Arthur Caplan, ed., *When Medicine Went Mad: Bioethics and the Holocaust* (Totowa, N.J.: Humana Press, 1992); see also George Annas and Michael Grodin, eds., *The Nazi Doctors and the Nuremberg Code* (New York: Oxford University Press, 1992). The Japanese wartime experiments are detailed in Sheldon Harris, *Factories of Death: Japanese Biological Warfare, 1932–45, and the American Cover-Up* (New York: Routledge, 1994). A postwar perspective on human experimentation in general is Andrew C. Ivy, "The History and Ethics of the Use of Human Subjects in Medical Experiments," *Science* 108 (1948): 1–5. Ivy was a medical advisor to the prosecution at the Nuremberg Trials. On the development of protections for human subjects, see David Rothman, *Strangers at the Bedside: A History of How Law and Bioethics Transformed Medical Decision Making* (New York: Basic Books, 1991); Ruth R. Faden and Tom L. Beauchamp, *A History and Theory of Informed Consent* (New York: Oxford University Press, 1986). Abuses in the 1960s were revealed by Henry Beecher, "Ethics and Clinical Research," *New England Journal of Medicine* 274 (1966): 1354–60, and Maurice Pappworth, *Human Guinea Pigs: Experimentation on Man* (London: Routledge and Kegan Paul, 1967). The Tuskegee syphilis experiments are analyzed in James H. Jones, *Bad Blood*, 2d ed. (New York: Free Press, 1991); see also Todd L. Savitt, "The Use of Blacks for Medical Experimentation and Demonstration in the Old South," *Journal of Southern History* 48 (1982): 331–48. Susan E. Lederer's *Subjected to Science* (Baltimore: Johns Hopkins University Press, 1995) discusses human experimentation in the United States before World War II. Lawrence K. Altman's *Who Goes First?* (Berkeley: University of California Press, 1998) surveys self-experimentation. Peter Singer's *Animal Liberation,* 2d ed. (New York: New York Review of Books, 1990) and Tom Regan's *The Case for Animal Rights* (Berkeley: University of California Press, 1983) remain the touchstones of the new movement for animal protection. Mary Midgley's *Animals and Why They Matter* (Athens: University of Georgia Press, 1984) is an important contribution to the debate, as is Andrew Rowan's *Of Mice, Models, and Men* (Albany: State University of New York Press, 1984). A good account of the issues around cloning is Gina Kolata, *Clone: The Road to Dolly and the Path Ahead* (New York: William Morrow, 1998); on the human genome project, see Matt Ridley, *Genome: The Autobiography of a Species in 23 Chapters* (New York: HarperCollins, 2000). The latest of a huge number of books published in the last two decades on animal experimentation is Deborah Rudacille, *The Scalpel and the Butterfly: The War between Animal Research and Animal Protection* (New York: Farrar, Straus, and Giroux, 2000). A better guide to modern issues is F. Barbara Orlans, *In the Name of Science* (New York: Oxford University Press, 1993); see also Jane A. Smith and Kenneth M. Boyd, eds., *Lives in the Balance: The Ethics of Using Animals in Biomedical Research* (Oxford: Oxford University Press, 1991).

Index

Made in the USA
Las Vegas, NV
12 October 2021